AF356655

Advances of Italian Machine Design

Advances of Italian Machine Design

Editors

Marco Ceccarelli
Giuseppe Carbone

MDPI • Basel • Beijing • Wuhan • Barcelona • Belgrade • Manchester • Tokyo • Cluj • Tianjin

Editors
Marco Ceccarelli   Giuseppe Carbone
University of Rome Tor Vergata   University of Calabria
Italy   Italy

Editorial Office
MDPI
St. Alban-Anlage 66
4052 Basel, Switzerland

This is a reprint of articles from the Special Issue published online in the open access journal *Machines* (ISSN 2075-1702) (available at: https://www.mdpi.com/journal/machines/special_issues/italian_machines).

For citation purposes, cite each article independently as indicated on the article page online and as indicated below:

LastName, A.A.; LastName, B.B.; LastName, C.C. Article Title. *Journal Name* **Year**, *Volume Number*, Page Range.

ISBN 978-3-0365-0906-8 (Hbk)
ISBN 978-3-0365-0907-5 (PDF)

Contents

About the Editors . . . . **vii**

Marco Ceccarelli and Giuseppe Carbone
Advances of Italian Machine Design
Reprinted from: *Machines* **2019**, 7, 61, doi:10.3390/machines7030061 . . . . **1**

Walter Franco, Carlo Ferraresi and Roberto Revelli
Functional Analysis of Piedmont (Italy) Ancient Water Mills Aimed at Their Recovery
or Reconversion
Reprinted from: *Machines* **2019**, 7, 32, doi:10.3390/machines7020032 . . . . **5**

Paolo Boscariol, Giovanni Boschetti, Aldo Dalla Via, Nicola De Rossi, Mauro Neri,
Ilaria Palomba, Dario Richiedei, Claudio Ronco and Alberto Trevisani
Description and In-Vitro Test Results of a New Wearable/Portable Device for Extracorporeal
Blood Ultrafiltration
Reprinted from: *Machines* **2019**, 7, 37, doi:10.3390/machines7020037 . . . . **25**

Daniela Maffiodo and Terenziano Raparelli
Flexible Fingers Based on Shape Memory Alloy Actuated Modules
Reprinted from: *Machines* **2019**, 7, 40, doi:10.3390/machines7020040 . . . . **43**

Antonio Acinapura, Gionata Fragomeni, Pasquale Francesco Greco, Domenico Mundo,
Giuseppe Carbone and Guido Danieli
Design and Prototyping of Miniaturized Straight Bevel Gears for Biomedical Applications
Reprinted from: *Machines* **2019**, 7, 38, doi:10.3390/machines7020038 . . . . **59**

Giuseppe Quaglia, Elvio Bonisoli and Paride Cavallone
The Design of a New Manual Wheelchair for Sport
Reprinted from: *Machines* **2019**, 7, 31, doi:10.3390/machines7020031 . . . . **75**

Giovanni Breglio, Andrea Irace, Lorenzo Pugliese, Michele Riccio, Michele Russo,
Salvatore Strano and Mario Terzo
Development and Testing of a Low-Cost Wireless Monitoring System for an Intelligent Tire
Reprinted from: *Machines* **2019**, 7, 49, doi:10.3390/machines7030049 . . . . **89**

Renato Brancati, Giandomenico Di Massa and Stefano Pagano
Investigation on the Mechanical Properties of MRE Compounds
Reprinted from: *Machines* **2019**, 7, 36, doi:10.3390/machines7020036 . . . . **99**

Claudia Aide González-Cruz and Marco Ceccarelli
Experimental Characterization of the Coupling Stage of a Two-Stage Planetary Gearbox in
Variable Operational Conditions
Reprinted from: *Machines* **2019**, 7, 45, doi:10.3390/machines7020045 . . . . **113**

İbrahimcan Görgülü, Mehmet İsmet Can Dede
A New Stiffness Performance Index: Volumetric Isotropy Index
Reprinted from: *Machines* **2019**, 7, 44, doi:10.3390/machines7020044 . . . . **125**

Nicola Secciani and Matteo Bianchi and Alessandro Ridolfi and Federica Vannetti and Yary Volpe and Lapo Governi and Benedetto Allotta
Tailor-Made Hand Exoskeletons at the University of Florence: From Kinematics to
Mechatronic Design
Reprinted from: *Machines* **2019**, *7*, 22, doi:10.3390/machines7020022 . . . . **141**

Kuatbay Bissembayev, Assylbek Jomartov, Amandyk Tuleshov and Tolegen Dikambay
Analysis of the Oscillating Motion of a Solid Body on Vibrating Bearers
Reprinted from: *Machines* **2019**, *7*, 58, doi:10.3390/machines7030058 . . . . **159**

Enrico Ciulli and Paola Forte
Nonlinear Response of Tilting Pad Journal Bearings to Harmonic Excitation
Reprinted from: *Machines* **2019**, *7*, 43, doi:10.3390/machines7020043 . . . . **181**

About the Editors

Marco Ceccarelli received PhD degree in 1987 at University La Sapienza of Rome. He is Professor at University of Rome Tor Vergata. He is Director of LARM2: Laboratory of Robot Mechatronics. He is Scientific Editor of Springer Book Series on History of MMS and on Mechanism and Machine Science, and Associate Editor of several Journals. He wrote the book Fundamentals of Mechanics of Robotic Manipulation. He co-authored a book on Mechanisms Design in Spanish and a book A brief illustrated history of machines and mechanisms. He received Degree of Doctor Honoris Causa from foreigner Universities and ASME Historian Award. He is ASME fellow. He was elected Secretary-General of IFToMM in 2004–2007 and IFToMM President in 2008–2011 and 2016–2019. His research cover aspects of Mechanism Design, Mechanics and Design of Robots, and History of Mechanical Engineering. He is author/co-author of several books and papers in international proceedings and journals. See http://larmlaboratory.net.

Giuseppe Carbone received the M.S. cum laude and PhD degree at University of Cassino (Italy) where he has been a Key Member of LARM (Laboratory of Robotics and Mechatronics) for about 20 years. Since Dec. 2018 he is Associate Professor at DIMEG, University of Calabria, Italy. His research interests cover aspects of Mechanics of Manipulation and Grasp, Mechanics of Robots, Mechatronic Devices with more than 300 published papers, He also received more than 20 best paper awards, 10 best patents in international contests. Among others, he currently serves as Chair of the IFToMM TC Robotics and Mechatronics, Member of the Board of Directors for the Society of Bionics and Biomechanics, Treasurer of the IFToMM Italy Society. He has been invited to deliver invited Keynote speeches and seminars on his research activity at more than 30 International events. He serves in the editorial board of several international journals, including being Editor in Chief of ROBOTICA (Cambridge Univ. Press), MDPI Machines, MDPI Robotics, TE or AE of several international Journals including IEEE/ASME Transactions on Mechatronics, ASME Journal of Autonomous Vehicles and Systems, Advanced Robotic Systems.

Editorial

Advances of Italian Machine Design

Marco Ceccarelli [1,*] and Giuseppe Carbone [2]

[1]   Department of Industrial Engineering, University of Rome Tor Vergata, 00133 Rome, Italy
[2]   DIMEG, University of Calabria, 87036 Cosenza, Italy; carbone@unicas.it
*   Correspondence: marco.ceccarelli@uniroma2.it

Received: 9 September 2019; Accepted: 9 September 2019; Published: 11 September 2019

This Special Issue is aimed to promote and circulate recent developments and achievements in the field of Mechanism and Machine Science coming from the Italian community with international collaborations and ranging from theoretical contributions to experimental and practical applications.

It contains selected contributions that were accepted for presentation at the Second International Conference of IFToMM Italy, IFIT2018, that has been held in Cassino on 29 and 30 November 2018 [1]. This IFIT conference is the second event of a series that was established in 2016 by IFToMM Italy in Vicenza. IFIT was established to bring together researchers, industry professionals and students, from the Italian and the international community in an intimate, collegial and stimulating environment.

IFToMM Italy is one of the founding member organizations of IFToMM, the International Federation for the Promotion of Mechanism and Machine Sciences in 1969. Since then, the member organization IFToMM Italy has been active with contributions at national and international levels, and in 2014 IFToMM Italy was legally established as the Italian IFToMM society.

This Special Issue includes papers belonging to a broad range of disciplines, such as the history of MMS (Mechanism and Machine Science), kinematics of mechanisms, transmissions, vehicle dynamics, transportation machinery, bearings, vibrations, design of robots, grasping, exoskeleton designs, medical devices and service systems. These contributions have been selected from among the 57 papers that were presented at IFIT 2018 conference [1] to have extended revised versions of the presented works. Most of them are those granted award recognition in one of the three IFToMM categories of research, applications, and student. These papers were evaluated again with a blind peer-review process to confirm the high quality of the works.

Three hand exoskeletons versions are presented in paper [2], looking at the mechanical design and control for a compact and lightweight solution that has been tested in a real-use scenario in the rehabilitation and assistance fields. Paper [3] presents an innovative system of propulsion for wheelchairs with details inspired by a rowing gesture, with possible application in everyday life and sport wheelchairs for speed races. In paper [4], the mechanical architecture of an old water mill is analyzed by means of few typical examples of an old water mill of the Piemonte region, in the northwest of Italy, looking at the functional details of various mechanisms for restoring them from the perspective of a renewed high quality production, or reconverting them in mini-plants for the production of electricity. Paper [5] discusses an experimental investigation on magneto-rheological elastomers with the aim of adopting these materials as vibration isolators. Paper [6] describes the design of Rene Artificiale Portatile (artificial portable kidney), a novel wearable and portable device for extracorporeal blood ultrafiltration, capable of providing remote treatment of fluid overload in patients with kidney diseases and/or congestive heart failure. Paper [7] presents a semi-automated design algorithm for computing straight bevel gear involute profiles based on the Tredgold approximation method. A specific case study of application of design results is discussed referring to the profiles of two straight bevel gears in a biomedical application for a new laparoscopic robotic system. Paper [8] presents finger designs consisting of a body with a flexible central rod and three longitudinally positioned shape memory alloy wires. This paper introduces a mathematical model for the design and discusses results of experimental tests on three prototypes different materials. Paper [9] deals with non-linear effects in tilting pad journal bearings with test results of an experimental identification procedure. Paper [10] proposes a new index for a precise calculation of a manipulator's stiffness isotropy with a numerical validation through an evaluation of an R-CUBE manipulator design. Paper [11] presents an experimental characterization of a two-stage planetary gearbox as function of the attached load and the internal friction forces between gears. Paper [12] deals with the use of cost-effective flex and polyvinylidene fluoride strain sensors to estimate some dynamic tire features in free-rolling and real working conditions, with a solution combining a microcontroller-based sensing with a wireless data transmission system. Paper [13] analyzes the oscillation of a solid body with rolling bearers through a model of highly nonlinear differential equations to design a vibro-protected solution.

We would like to thank the members of the Scientific Committees for strong support for the success of IFIT 2018:

Alessandro Gasparetto (University of Udine) Chair
Nicola Pio Belfiore (University of Roma)
Massimo Callegari (Polytechnical University of Marche)
Roberto Caracciolo (University of Padova)
Giuseppe Carbone (University of Calabria)
Marco Ceccarelli (University of Rome Tor Vergata)
Enrico Ciulli (University of Pisa)
Raffaele di Gregorio (University of Ferrara)
Pietro Fanghella (University of Genova)
Andrea Manuello Bertetto (Politecnico di Torino)
Arcangelo Messina (University of Salento)
Domenico Mundo (University of Calabria)
Vincenzo Niola (University of Napoli)
Paolo Pennacchi (Politecnico di Milano)
Giuseppe Quaglia (Politecnico di Torino)
Rosario Sinatra (University of Catania)
Alberto Trevisani (University of Padova)

The guest editors of this Special Issue thank the authors and reviewers for their efforts and time spent in the valuable scientific contributions and useful feedbacks that have confirmed the high scientific quality of the IFIT 2018 papers.

Conflicts of Interest: The authors declare no conflict of interest.

References

1. Carbone, G.; Gasparetto, A. (Eds.) *Advances in Italian Mechanism Science. Proceedings of the Second International Conference of IFToMM Italy*; Springer: Cham, Switzerland, 2019; Volume 68.
2. Secciani, N.; Bianchi, M.; Ridolfi, A.; Vannetti, F.; Volpe, Y.; Governi, L.; Bianchini, M.; Allotta, B. Tailor-Made Hand Exoskeletons at the University of Florence: From Kinematics to Mechatronic Design. *Machines* **2019**, *7*, 22. [CrossRef]
3. Quaglia, G.; Bonisoli, E.; Cavallone, P. The Design of a New Manual Wheelchair for Sport. *Machines* **2019**, *7*, 31. [CrossRef]
4. Franco, W.; Ferraresi, C.; Revelli, R. Functional Analysis of Piedmont (Italy) Ancient Water Mills Aimed at Their Recovery or Reconversion. *Machines* **2019**, *7*, 32. [CrossRef]
5. Brancati, R.; Di Massa, G.; Pagano, S. Investigation on the Mechanical Properties of MRE Compounds. *Machines* **2019**, *7*, 36. [CrossRef]

6.   Boscariol, P.; Boschetti, G.; Dalla Via, A.; De Rossi, N.; Neri, M.; Palomba, I.; Richiedei, D.; Ronco, C.; Trevisani, A. Description and In-Vitro Test Results of a New Wearable/Portable Device for Extracorporeal Blood Ultrafiltration. *Machines* **2019**, *7*, 37. [CrossRef]

7.   Acinapura, A.; Fragomeni, G.; Greco, P.; Mundo, D.; Carbone, G.; Danieli, G. Design and Prototyping of Miniaturized Straight Bevel Gears for Biomedical Applications. *Machines* **2019**, *7*, 38. [CrossRef]

8.   Maffiodo, D.; Raparelli, T. Flexible Fingers Based on Shape Memory Alloy Actuated Modules. *Machines* **2019**, *7*, 40. [CrossRef]

9.   Ciulli, E.; Forte, P. Nonlinear Response of Tilting Pad Journal Bearings to Harmonic Excitation. *Machines* **2019**, *7*, 43. [CrossRef]

10.   Görgülü, İ.; Dede, M. A New Stiffness Performance Index: Volumetric Isotropy Index. *Machines* **2019**, *7*, 44. [CrossRef]

11.   González-Cruz, C.; Ceccarelli, M. Experimental Characterization of the Coupling Stage of a Two-Stage Planetary Gearbox in Variable Operational Conditions. *Machines* **2019**, *7*, 45. [CrossRef]

12.   Breglio, G.; Irace, A.; Pugliese, L.; Riccio, M.; Russo, M.; Strano, S.; Terzo, M. Development and Testing of a Low-Cost Wireless Monitoring System for an Intelligent Tire. *Machines* **2019**, *7*, 49. [CrossRef]

13.   Assylbek, K.; Jomartov, A.; Tuleshov, A.; Dikambay, T. Analysis of the Oscillating Motion of a Solid Body on Vibrating Bearers. *Machines* **2019**, *7*, 58. [CrossRef]

Article

Functional Analysis of Piedmont (Italy) Ancient Water Mills Aimed at Their Recovery or Reconversion [†]

Walter Franco [1,*], **Carlo Ferraresi** [1] **and Roberto Revelli** [2]

[1] Politecnico di Torino, Department of Mechanical and Aerospace Engineering- DIMEAS, 10129 Torino, Italy; carlo.ferraresi@polito.it

[2] Politecnico di Torino, Department of Environment, Land and Infrastructure Engineering- DIATI, 10129 Torino, Italy; roberto.revelli@polito.it

[*] Correspondence: walter.franco@polito.it; Tel.: +39-011-090-3348

[†] This paper is an extension version of the conference paper: "Power Transmission and Mechanisms of an Old Water Mill"; Walter Franco, Carlo Ferraresi, and Roberto Revelli in Proceedings of the Second International Conference of IFToMM Italy, Cassino, Italy, 29–30 November 2018.

Received: 10 April 2019; Accepted: 10 May 2019; Published: 13 May 2019

Abstract: Since ancient times and for hundreds of years, grain mills, hammers, sawmills, spinning mills, and hemp rollers have been powered by water wheels. In the nineteenth century there were hundreds of thousands of mills in all of Europe. It is an enormous historical and cultural heritage of inestimable value, which is for the most part, abandoned today. Recently, there is a renewed interest in their reuse, both for their widespread diffusion in the territory and for the excellent environmental integration and intrinsic sustainability. Even when, for economic reasons, their recovery for the original tasks is not suitable, the conversion into mini plants for the production of electricity can be advantageous. In the paper, analyzing some typical examples of the old water mill of the Piemonte region, in North-West of Italy, the mechanical architecture of old water mill, from water wheels to millstones, is described and the functional details of various mechanisms are provided. In fact, by knowing only the specifics of the ancient mills, it is possible to enhance their potential and restore them from the perspective of a renewed high quality production, or reconvert them in mini-plants for the production of electricity.

Keywords: water wheel; grain water mill; wooden teeth gear; history of mechanism and machine science; micro-hydro; renewable energy

---

1. Introduction

The history of water mills matches the development of prime movers. In general, the aim is to make more energy available for a community, in particular more concentrated in space and time. In fact, the development of a new engine allows an increase in the level of production of a society.

Singer et al. [1] have identified different stages in the evolution of the prime movers. In the history of humanity, the first traditional available energy has been the human one. Using muscles over motor, the human body can express a continuous power close to 100 W, that could be used for limited production activity [2,3]. A first improvement, in terms of total available energy and concentrated power, was obtained using animals as a power source. An animal can continuously develop a power of about 1 W per kilogram, so that a medium ox expresses a power of approximately 500 W, five times that generated by a single man. The second substantial step in the progress of prime movers was, precisely, the development of water wheels. To understand the impact of this innovation, one must consider that, at the end of the Roman Empire, water wheels were able to develop about 45 kW of power, more than four hundred and fifty people [1].

There are two main types of mills: The Greek or Norse and the Vitruvian. The Greek mill originated probably in the mountainous areas of the Near East. The first traces are narrated by Strabo that, in 65 B.C., tells us about a mill built by Mithridates, the king of Pontus, near his palace in Cabeira. This kind of mill has spread both to the east and to the west, and in the fourth century is present throughout Europe. It presents a very simple mechanical architecture: The horizontal wheel is directly coupled to the load through a vertical shaft, supported by an axial bearing cooled by the water flow. The horizontal water wheel is appropriate for the low flow rate and high water velocity, typical of mountain rivers. The mean rotational velocity is about 15–20 rpm. The conversion into mechanical energy occurs with low efficiency. The horizontal water wheel was normally used as the prime mover of small grain mills, suitable for a modest production of flour of the order of a few quintals a day, typical of highland family farming. In some cases, it was also used for running hammers of blacksmith's workshops and hemp mills [4], devoted to roll the hemp fibers in order to break down their natural coating and make it pliable enough for spinning and weaving (Figure 1).

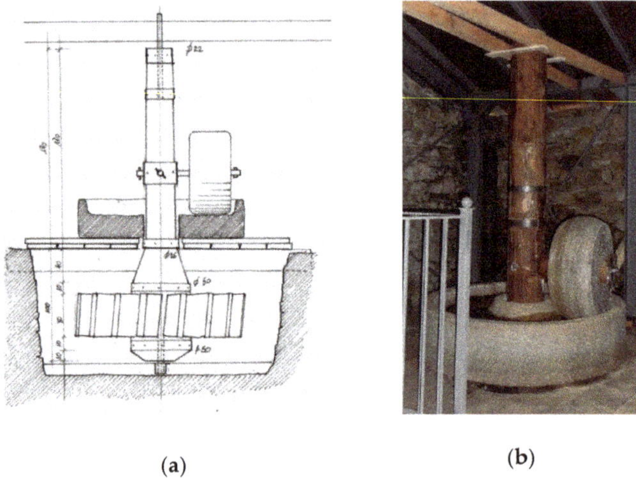

(a)   (b)

Figure 1. Hemp roller mill of Combe, Celle Macra (Italy). (**a**) Scheme of the horizontal wheel keyed to the vertical shaft; (**b**) the mill after the renovation. Courtesy of Arch. Roberto Olivero.

The other kind of mill is the Roman Vitruvian, characterized by a more efficient vertical water wheel keyed to a horizontal shaft. As the power take off often requires a vertical shaft, a more complex power transmission, able to change the direction of the axis of the shafts and, if necessary, to multiply its velocity, is required. The problem was solved with the invention of crown and lantern gears, which, in spite of the augmented complexity, allowed to increase significantly the overall efficiency and productivity of the mill [1].

Initially, in the Roman Era, the spread of the mills has been very slow, both due to the availability of cheap labor, and to the difficulty in regulating the regime of the rivers of the Mediterranean area, characterized by strong seasonal fluctuations in flow. In fact, all the oldest mills found in archaeological excavations, Venafro, Barbegal, Tournus, Rome and Athens, were fed by an aqueduct. In the Middle Ages, decreasing the availability of workforce, water mills spread considerably, so that in the nineteenth century there were more than 60,000 wheels in France, 30,000 in England and 33,500 in Germany. Water wheels were employed mainly for grain milling, but also for running forge hammers, bellows, mortars, fullers, and saws [5,6]. In all civilizations of the past, water wheels (noria) were also used to elevate water [7]. The peak of the scientific improvement in water wheels design is probably the introduction, in the middle of the nineteenth century, of the Sagebien and Zuppinger water wheels. They replaced the previous models with a considerable efficiency improvement. In the same period, we also see a flowering of scientific works, mostly experimental or concerning design suggestions [8].

The water wheels use and interest declined only with the construction of big hydroelectric plants and the advent of modern turbines (Francis, Pelton and Kaplan). In the first decades of the twentieth century, dozens of dams were built: The power produced by water wheels cannot compete with the megawatt produced by such power plants.

Today, the numerous old mills, although for the most part disused, represent a huge and precious heritage, which, beyond its historical and cultural value, has a remarkable potential. Old water mills, in fact, could be restored both for the direct actuation of machines, in the case of museum exhibitions or small high quality local productions (millstones, blacksmith's hammers, etc.), and for the conversion into mini-plants for the production of electricity.

On the other hand, water wheels and old mills have significant advantages:

(i)   they do not require the construction of new masonry, or water derivation channels;
(ii)   they already have a water use concession, often still ongoing;
(iii)   they are perfectly integrated into the environment;
(iv)   they are welcomed by the local population, being perceived as not impacting, beautiful, related to memory;
(v)   they are often widespread in isolated and underserved internal areas, such as small mountain villages;
(vi)   they use renewable energy.

Precisely because of the historical value of the water mills, Rojas-Sola et al. [9] modeled a mill using SolidWorks CAD, and generated a computer animation of the production process as a subject of a course in the history of technology.

Rojas-Sola et al. in addition studied, from a hydraulic and mechanical point of view, typical Spanish watermills, with horizontal water wheels. The aim of the work was graphically modeling the mill, in order to understand its dimension and to be able to make a virtual recreation of its functioning, and secondly to obtain the principal technical parameters, such as the supply flow rate and the power of the water wheel [10,11].

Pujol et al. conducted a detailed analysis of the performance of ancient Spanish horizontal water wheels [12] and the study of the implications of several technological innovations applied to the old classical horizontal waterwheels implemented in Gaserans (North-East Spain) [13].

The present work aims to describe the mechanical architecture of typical Piedmont (Italy) Vitruvian mills, and to analyze in detail the functioning of the water wheel, the mechanical power transmission and the regulation mechanisms. First, the research was carried out for a historical interest, but also in the perspective of recovering the original use of old mills, or for their reconversion into micro hydropower generators for electricity production. The economic sustainability of a restructuring operation of an old mill depends in fact on the state of preservation of the site, on the type of water wheel installed, on the condition of the mechanical transmissions, and, in summary, on the overall efficiency of the whole machine, from the wheel to the load.

The paper is organized as labeled below. First, different types of vertical water wheels are described. Then, the power transmission, between the wheel and the millstones, is analyzed from a functional point of view. Thereafter, the regulation systems and the mechanisms of the mill are presented. Concluding, the usefulness of the development of a device devoted to the experimental characterization of the efficiency of the mill transmission, from the water wheel to the load, is discussed.

For this purpose, two old mills, the Riviera mill and the Forno mill, are analyzed in detail.

The Riviera mill is a virtuous example of the restoration and reactivation of a grain mill for small productions of high quality organic flours in the short food supply chain [14]. The Riviera mill (Figure 2) is located in Dronero (North-West of Italy—44°28′ N and 7°21′ E) and dated back to the XV century [4,15]. It was renewed and extended several times [4]. In 1811, a first building extension request was made to realize a sawmill and add a water wheel. In 1813, it is attested that the mill has three millstones. In 1819, the addition of a new wheel is authorized. The main extension is carried out in 1859, as evidenced by an inscription on the entrance porch. In 1919 the mill, equipped with three millstones, is able to produce 10 quintals of flour per day. The mill was used for grinding wheat and maize corn until the 1970s, when it was abandoned. In 2002, the Cavanna family bought and renovated the Riviera mill, and re-started the milling of flours with natural stones.

The Forno family is the owner of a mill that produced flour, corn by-products and animal feed until 2011. The Forno mill (Figure 3) is located in Verolengo, in the Metropolitan City of Turin (North-West of Italy—45°11′ N and 7°58′ E). From a deep documentary analysis, it is possible to set the mill's building between 1874 and 1878 and, presumably, it was built on the ruins of an older abusive mill. The first presence of the Forno mill is reported in a military map dated 1882. From a more recent note, dated 1933, it is possible to deduce the start of the activities in 1879 and the characteristics of the permit to derive water from the "Roggia dei Mulini". In 1946, the productive apparatus were enlarged and the original millstones replaced. The production definitely stopped in 2011. The mill is actually under consideration for a possible inclusion in a proposal for an ecomuseum.

(a)   (b)

Figure 2. The Riviera mill: (**a**) Before the renovation (2004); (**b**) today.

Figure 3. Forno mill: The water inlet and the two water wheels.

2. The Water Wheels

Gravity water wheels (GWWs) are members of the group of machines that are used to convert hydro energy into mechanical energy. In general, we can categorize them according to the exploited energy transformation. Action turbines exploit the flow momentum (i.e., kinetic energy), reaction turbines convert both the flow momentum and the water pressure (they are installed in closed pipes) and hydrostatic pressure converters (HPCs) exploit water hydrostatic force. In a HPC, the hydrostatic force is generated by the water weight contained inside the machine buckets. The stream and vertical axis water wheels, Pelton, Turgo and Cross Flow turbines belong to the first group. In the second group the Kaplan and Francis turbines are present, while the last group includes gravity water wheels (GWWs) and Archimedes screws. GWWs rotate around a horizontal axis while in an Archimedes screw, the axle is inclined on the horizontal of about 22° to 35°.

Gravity water wheels are generally divided in several types, depending on the water entry point (Figure 4). In an overshot GWW, the water enters from the top. In breastshot WWs, the water enters from the upstream side and, depending on the water level position with respect to the rotational axis they can also be divided in high, middle and low breastshot GWWs. The low breastshot WW is often classified as an undershot GWW. The overshot GWWs (Figure 4a) rotate in a clockwise direction, on the opposite breastshot and undershot GWWs rotate in a counterclockwise direction (Figure 4b–d). Sagebien and Zuppinger GWWs differ, in particular, for the blades geometry: Sagebien and Zuppinger GWWs the curved blades optimize the inflow and outflow power losses, respectively.

Operatively (Table 1), WWs have a maximum efficiency between 70% and 90%. The typical exploitable head is 1–6 m, flow rate lower than 1 m^3/s and power less than 50 kW. The Riviera mill and the Forno mill consist of overshot and breastshot GWWs, respectively, with horizontal rotating axles.

Table 1. Typical characteristic data of gravity water wheels (GWWs). Cost refers to energy production (see [16] for details).

Type	Head m	Flow Rate m^3/s	Max Eff. %	Cost €/kW	Payback Time Years
Overshot	3–6	0.2–0.4	80–85	3900–4300	7.5–8.5
Breastshot	1–4	0.3–1	70–85	4000–7000	8–12
Undershot	<1.5	1	70–85	6900–8700	12–17

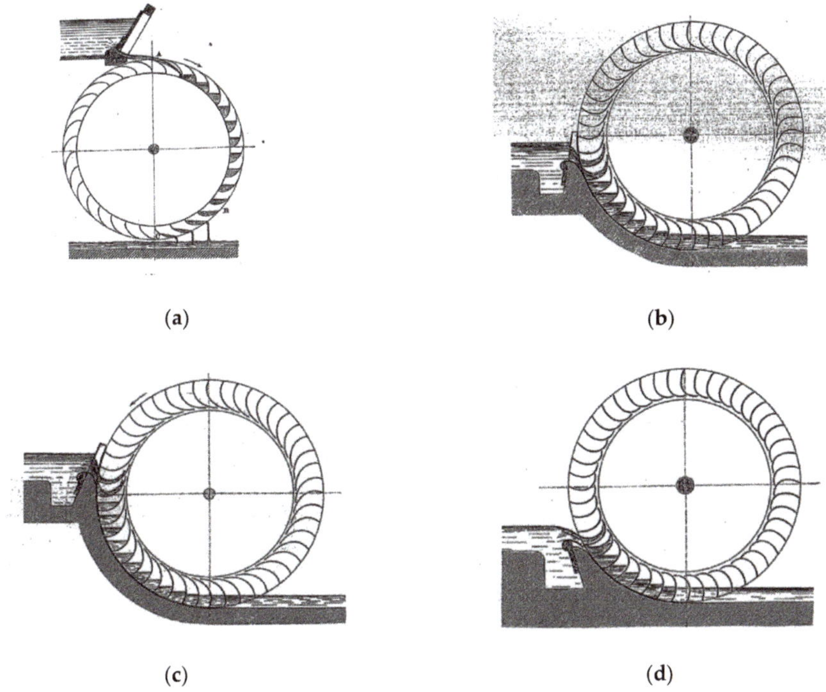

(a)

(b)

(c)

(d)

Figure 4. *Cont.*

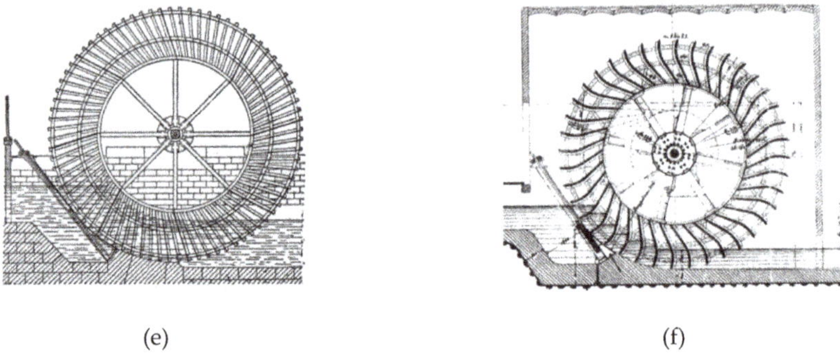

(e)   (f)

Figure 4. Types of gravity water wheels: (**a**) Overshot; (**b**) low breastshot; (**c**) high breastshot; (**d**) undershot; (**e**) Sagebien; (**f**) Zuppinger. Adapted from [17–19].

The water wheel shaft is normally supported by two plain bearing, lubricated by grease or oil, mounted across the canal on a metal frame (Figure 5b) or a brick structure (Figure 6b).

(a)   (b)

Figure 5. The overshot water wheels of the Riviera mill: (**a**) Before the renovation (Courtesy of Arch. Roberto Olivero, 2004); (**b**) after the renovation.

(a)   (b)

Figure 6. *Cont.*

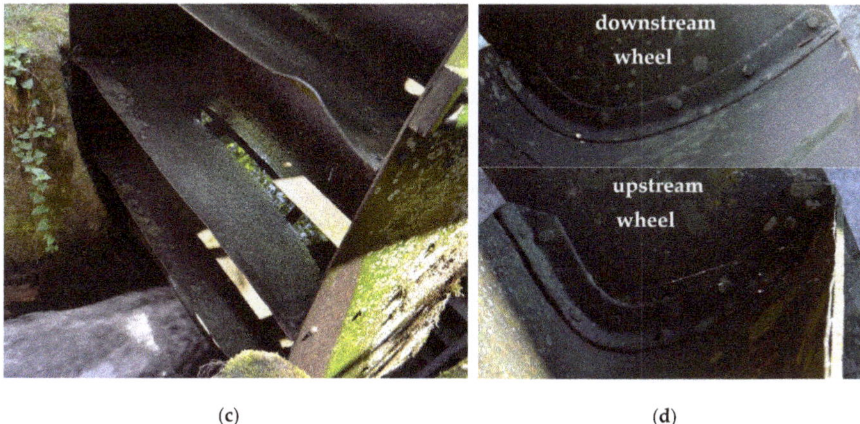

(c)   (d)

Figure 6. The low breastshot water wheels of the Forno mill: (a) Overall view; (b) side view and particular of the worn plain bearing; (c) water entry point; (d) shape of the blades.

The Riviera mill consists of two metallic WWs (Figures 2 and 5) and the water is supplied by the Maira River through the Bealera Comella. The wheels are 1.3 m wide and have a diameter of 3.0 m. They are an overshot type with 30 blades and rake angles equal to 60°. The Forno mill uses the water derived from the Dora Baltea River through a complex system of irrigation canals (Roggia Natta, Roggia del Veuchio, Roggia di Neirole). The system dates back to the 15thcentury and allowed to convey the granted flow rate in the Roggia dei Mulini. The two metallic water wheels (Figure 6) have a diameter of 4.0 m, width of 1.4 m and 32 blades.

A complete analysis of the efficiency, output power and power losses of the Riviera mill and Forno mill WWs has been performed through theoretical analyses, experimental measurements in a laboratory device and with the help of CFD numerical simulations [20–25]. The results have been also used for the check of ancient and popular past formulation [8,26].

For the sake of simplicity, we report here the main results of the hydraulic study, leaving it to the interested reader to deepen in the cited papers. The efficiency is related to water energy input from the upstream channel, power output and power losses and these latter occur in several parts of the installation

$$\eta = \frac{P_{out}}{P} = \frac{P - \sum losses}{P}$$

where η is the WW efficiency, P_{out} is the mechanical output power and $P = P_u - P_d$ is the difference between the water power upstream (P_u) and downstream (P_d) the WW, respectively. The losses depend on the WW type. In the overshot case (Riviera mill)

$$\sum losses = L_{inp} + L_t + L_g + L_{Q_u} + L_{Q_r}$$

where L_{inp} is the power loss occurring in the impact, L_t is the possible impact loss generated when the blades impact against the tailrace, L_g is the mechanical friction loss at the shaft supports, L_{Q_u} is the volumetric loss at the top of the wheel and L_{Q_r} is the volumetric loss during rotation. Similarly, in the breastshot WW (Forno mill)

$$\sum losses = L_{inp} + L_t + L_g + L_{Q_u} + L_c + L_Q + L_{bed} + L_h$$

where L_c are the hydraulic losses between the sluice gate and the WW, L_Q is the loss due to water that filters from the slits between the buckets and the channel, L_{bed} is the drag effect of water on the channel bed and L_h is the loss that occurs when the residual power of the water in the last bucket is lost in the tailrace. In [20,26] the power losses are related to the WW's geometric characteristics and the channel's hydraulic characteristics.

The theoretical, experimental and numerical analysis allowed to identify the reference results reported in Table 2. In particular, the Table shows the wheel efficiency and the relative importance of the power losses with the consequent suggestion for possible efficiency improvements [25,27].

Table 2. Reference values for the two WW (see [20,26] for details).

WW	Q	ω	P	T	η	L_c	L_{Q_u}	L_{inp}	L_Q	L_{bed}	L_t	L_h	L_g	L_{Q_r}
	m^3/s	rad/s	W	Nm	-	W	W	W	W	W	W	W	W	W
Riviera Mill	0.300	0.89	8500 [1]	9550	80%	-	85	680	-	-	~0	-	170	1950
Forno Mill	0.400	0.63	4790	7600	78%	380	96	480	530	15	~0	480 [2]	120	-

(1) Calculated with filling volume of the buckets 60%. (2) See [26] for the significance of the negative value.

3. The Power Transmissions

In a Vitruvian mill, the power transmission devices, included between the water wheel and the load, have a dual purpose: (i) To change the direction of the axis of the shafts, from horizontal and perpendicular to the main wall of the mill, to vertical, for example to rotate the runner stone in the grain mills; (ii) to multiply the angular velocity of the load shaft, in order to increase the productivity.

As an example, Figure 7 shows the scheme of the Riviera mill power transmission. An overview of the same power transmission is reported in Figure 8a. Each of the two water wheels can be connected to two runner stones through an independent power transmission system, but the single water wheel can drive only one millstone at a time. The two power transmission systems are analogous, therefore only the one operated by the downstream water wheel is described below in detail. The downstream water wheel is keyed to the shaft A, called the main shaft ($\Phi = 120$ mm), that crosses the mill wall. A first bevel gear multiplier (Figure 8b), with teeth number respectively $z_1 = 96$ and $z_2 = 40$, drives the shaft B, said lay shaft, ($\Phi = 78$ mm), arranged horizontally in parallel to the perimeter wall of the mill. The first gear of the bevel multiplier, keyed to the water wheel shaft, is called the pit wheel; the second gear is called the wallower. A second bevel gear train (Figure 9), with teeth number respectively $z_3 = 90$ and $z_4 = 34$, further multiplies the rotation speed of the output vertical shaft C, called the spindle shaft ($\Phi = 80$ mm), that actuates the runner stone one.

The overall gear ratio ω_C/ω_A is then equal to 6.35, whereby the nominal rotation speed of the millstone ω_C is about 60 rpm, while the rotation speed of the water wheel ω_A is about 10 rpm.

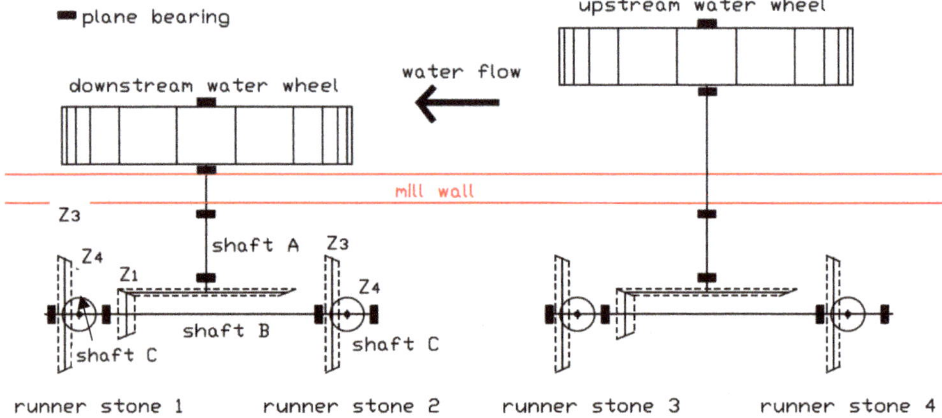

Figure 7. Plan view scheme of the Riviera mill power transmission.

(**a**)   (**b**)

Figure 8. The Riviera mill: (**a**) Power transmission overview; (**b**) the first bevel gear multiplier.

The rotation of the runner stone can be stopped by moving the bevel gear z_4 along the spindle vertical shaft C through a special lever (Figure 9b), until the contact between teeth z_3 and z_4 is interrupted.

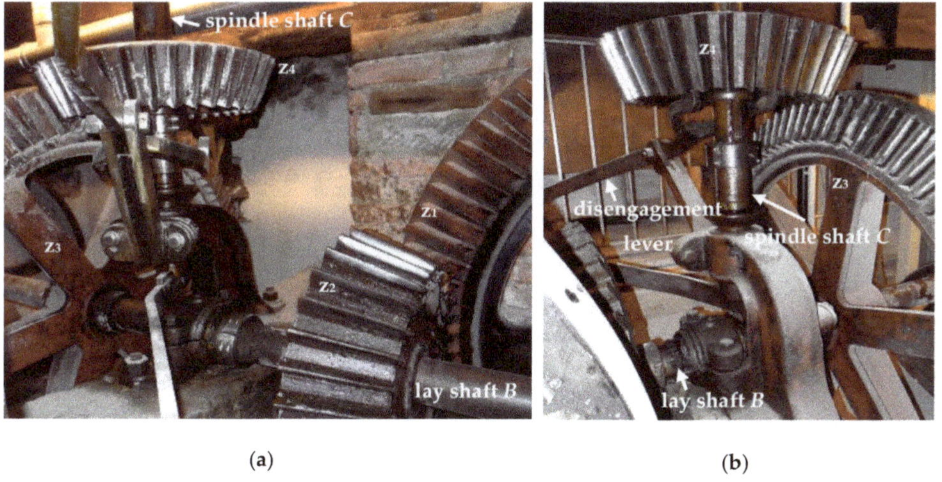

(**a**)   (**b**)

Figure 9. The Riviera mill power transmission: (**a**) Engagement of the runner stone; (**b**) disengagement of the runner stone.

The power transmission architecture described above is common to that of several old grain mills. It is found, for example, in the mechanical transmission of the upstream water wheel of the Forno mill that can drive two running stones as shown in the scheme of Figure 10.

Here again, the change in the direction of the shafts, and the multiplication of angular velocity of the runner stone are realized by two bevel gears (Figure 11), whose functional parameters are summarized in Table 3. The overall gear ratio ω_C/ω_A is equal to 6.57, almost equal to the overall speed ratio of the Riviera mill. The only difference consists in the plan distribution of the transmission, imposed by the conformation of the building. While in fact in the case of the Riviera mill, the wheel z_2 is keyed in the middle of the lay shaft B, between the two millstones (Figure 7), in the case of the Forno mill it is positioned at the right end of the lay shaft B (Figure 10).

Table 3. Functional parameters of the power transmissions.

Parameter	Riviera Mill	Forno Mill Stone Mill	Forno Mill Roller Mill
ω_W	10 rpm	8–12 rpm	8–12 rpm
z_1	96	64	96
z_2	40	32	31
z_3	90	92	-
z_4	34	28	-
ω_C/ω_A	6.35	6.57	≈12 (valued)
ω_C	63 rpm	52.5–78.8 rpm	96–144 rpm

Additionally, in the case of the Forno mill, the actuation of each runner stone can be stopped by operating a special lever, whose function is to move vertically the gear z_4, connected through a key to the spindle shaft C, until the contact between teeth z_3 and z_4 is interrupted (Figures 11b and 12). In this case, the disengagement lever is hinged to the wooden beam, which supports the floor on which the millstones are placed (stone floor).

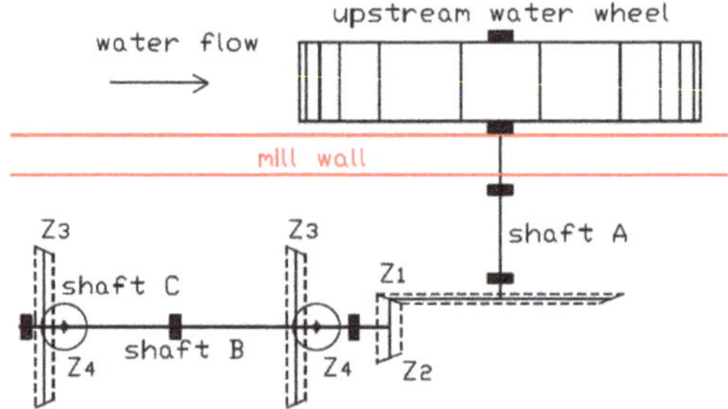

Figure 10. Plan view scheme of the Forno mill upstream water wheel.

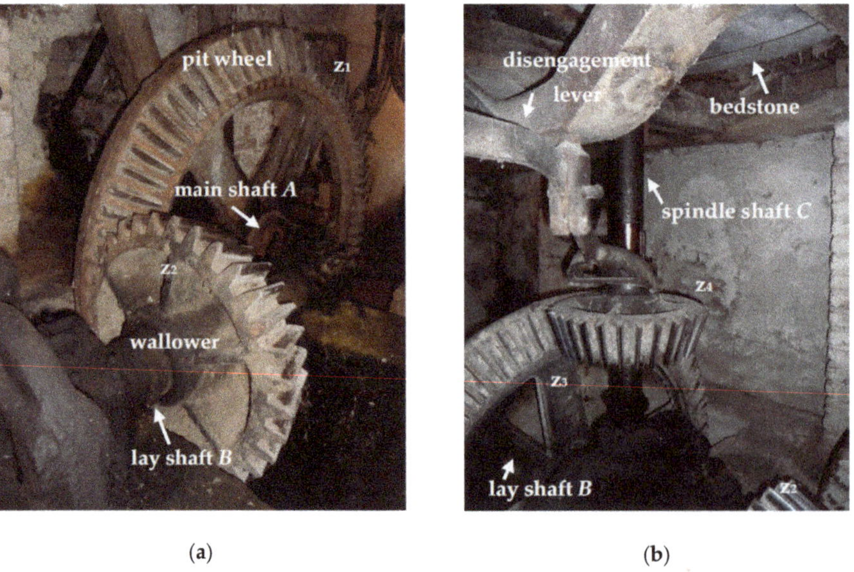

(**a**)

(**b**)

Figure 11. The Forno mill upstream water wheel power transmission: (**a**) The first bevel gear multiplier; (**b**) the second bevel gear multiplier.

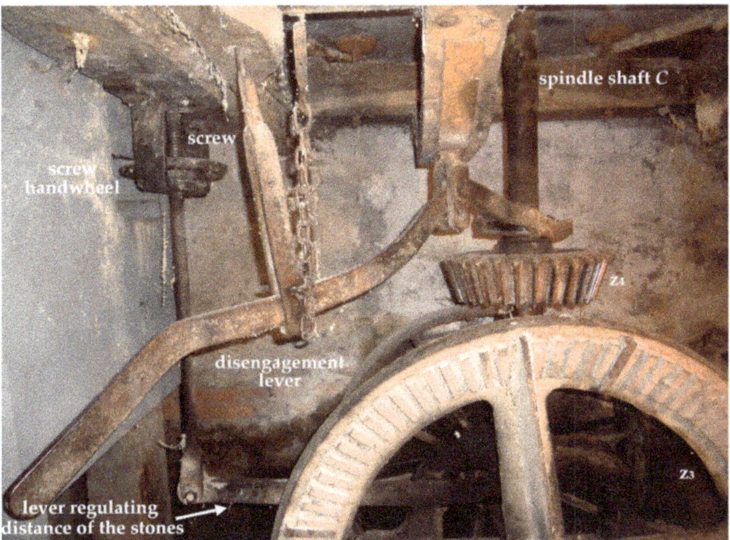

Figure 12. Disengagement of the runner stone and lever for regulation of the distance of the stones in the Forno mill.

The downstream water wheel of the Forno mill, instead, drives more recent roller mills. After the first speed multiplier (Figure 13a), consisting of the usual bevel gear pair (see Table 3: $z_1 = 96$, $z_2 = 31$; $z_1/z_2 = 3.09$) there is a gearbox consisting of a pair of straight teeth spur gears, that transmit the power to a third shaft C, now horizontal and parallel to the main building wall (Figure 13b). Although it was not possible to open the box of the multiplier, its multiplication ratio was evaluated to be about four, simply by estimating the radius of the two gears. It follows that the overall gear ratio ω_C/ω_A is about 12, almost twice the overall gear ratio of the oldest upstream wheel power transmission. Several pulleys with a diameter of about 550 mm are connected to the shaft C and transmit, by flat belts, the power to the roller mills of manufacture Amme Giesecke & Konegen, Braunschweig, Germany (Figure 14). To prevent accidents due to the use of non-protected belt drives, several old injury prevention sign are displayed in the mill (Figure 14c,d).

(**a**)   (**b**)

Figure 13. The Forno mill downstream water wheel power transmission: (**a**) The first bevel gear multiplier and the spur gearbox; (**b**) the pulleys actuating the roller mills.

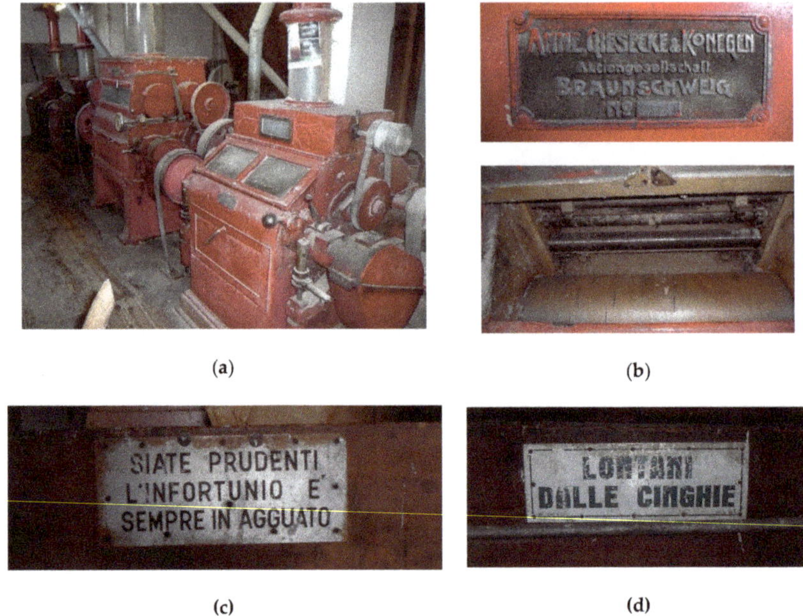

Figure 14. The Forno mill: (**a**) Amme Giesecke & Konegen roller mills; (**b**) particular of the label and inside of the mill; (**c**) and (**d**) old accident prevention signs: "Be prudent; the injury is always lurking" and "Far from the belts".

In all the bevel gear pairs, the teeth of the first gear are made of apple wood (Figure 15). The wooden teeth present the following advantages: (i) They can be easily manually produced with a simple band saw; (ii) they introduce a mechanical compliance in the transmission that absorbs the impulsive loads during the starting and stopping transient states; (iii) they reduce the noise generated by the mechanical transmission; (iv) they are prone to wear preserving the integrity of the metal teeth (Figure 15c,e); (v) they are easily replaceable once worn, being set in a rim, in which they are driven and keyed (Figure 15d,e).

On the other hand, both because the teeth do not have the proper shape and for the considerable wood friction coefficient, the efficiency of the mechanical transmission with wooden teeth mortise wheel is certainly low. This must be carefully assessed in the case of reconversion of an ancient mill for the production of electricity, and measured as discussed in conclusions.

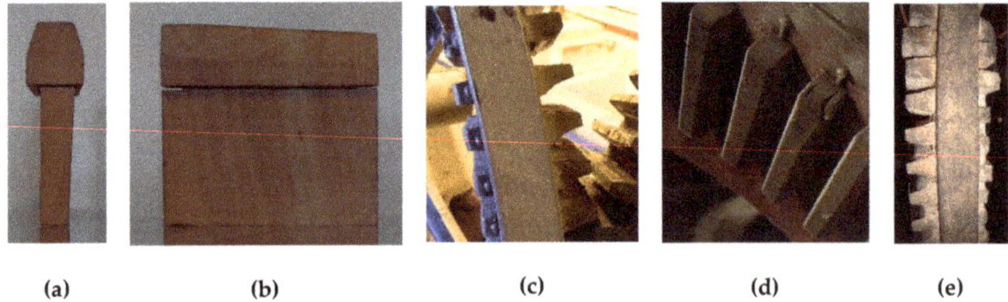

Figure 15. The mortise gear (Riviera): (**a**) Side view; (**b**) frontal view; (**c**) worn teeth; (**d**) particular of the wooden teeth mounting on the wheel ring; (**e**) wear and teeth mounting (Forno mill).

4. Regulation Devices

The Riviera mill vertical shaft *C* (spindle) supports and moves the runner stone, on the top of the fixed one, called bedstone (Figure 16).

The connection between the shaft C and the runner stone is made with the coupling of the Figure 17a,b, in which it is also possible to see the radial plain bearing of the shaft C integral with the bedstone, made of Teflon after the renovation of the Riviera mill. Figure 17c shows, from below, the original same plain bearing used in the Forno mill.

The vertical load of the runner stone, whose mass is around 7500 kg, is supported by an oil-immersed plain axial bearing (Figure 18a). In order to set the optimal distance between the mill stones, for grinding different types of cereals or to produce a finer or coarser flour, the axial bearing may be raised or lowered by a lever (bridge tree) operated remotely with a lead screw mechanism called the tentering screw (Figures 12, 16 and 18b).

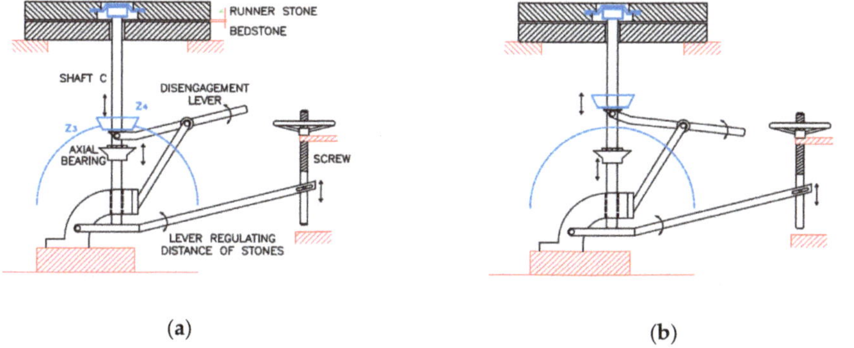

Figure 16. Mechanical scheme of the runner stone engagement lever and of the stones distance regulation mechanism (Riviera mill): (**a**) Runner stone engaged; (**b**) runner stone disengaged.

Figure 17. *Cont.*

(c)   (d)

Figure 17. Coupling between the shaft C and runner stone: (**a**) Shaft C (Riviera mill); (**b**) runner stone (Riviera mill); (**c**) shaft C (Forno mill); (**b**) runner stone (Forno mill).

(a)   (b)

Figure 18. Stone distance regulation mechanism of the Riviera mill: (**a**) Axial bearing; (**b**) lead screw mechanism.

Both the runner stone and the bedstone are contained in a wooden cover case called tun or vat (Figures 17d and 19a), positioned on the wooden beams of the stone floor, with the function of collecting the flour and conveying it to the plan sifter. The grain, that has to be ground, is loaded into a hopper located at the top of the tun. Due to gravity, the grain falls on an inclined plane (shoe), shaken laterally by a cam (damsel) actuated by a shaft keyed with the runner stone (Figure 20), so that the oscillation frequency of the inclined plane depends on the rotation speed of the runner stone. This means that the flow of the grain that feed the central hole of the runner stone grows by increasing the grinding rotation velocity. In addition, the grain flow can be adjusted by changing the slope of the inclined plane (Figure 20) by regulating the length of the string (Figures 20 and 21).

(a)   (b)

Figure 19. Stone floor of Riviera mill: (**a**) Tun and hopper; (**b**) lifted runner stone.

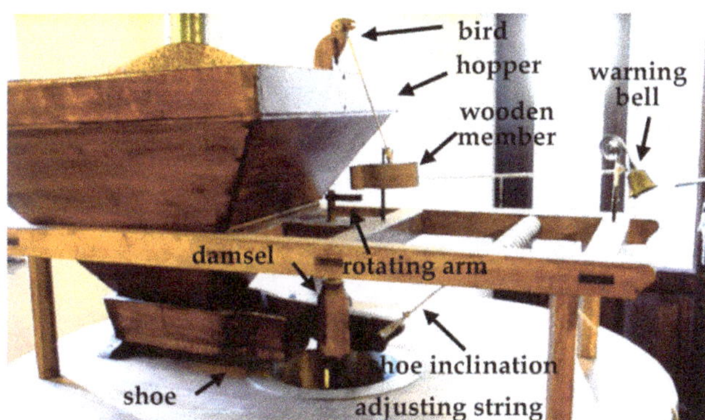

Figure 20. Grain flow adjustment mechanism and hopper emptying alarm (Riviera mill).

To avoid damages to the millstones, it is necessary to ensure that the hopper never completely empties and the grain feeding is never interrupted. With this aim, a curious warning system has been designed. As long as there is a sufficient amount of cereal in the hopper, a wooden bird (Figure 20) remains in a standing position, held by a rope immersed in the grain (not visible in the figure). In this condition, the wooden member, connected to the warning bell by a string, is not impacted by the rotating arm keyed to the same shaft of the cam shaker. If the hopper gets to low, the rope is released, the bird rotates around a pin, and it puts down the head as if it was pecking. In this way, the wooden element connected to the warning bell goes down, and interferes with the rotating arm that, by striking it, causes the alarm bell to ring during each revolution of the mill stone.

Figure 21. Feeding the millstones.

Finally, to obtain good quality flour, it is also important to adjust the grinding speed, i.e., the water wheel supplying flow rate. In the Riviera mill, a regulating lever, positioned inside the build at the stone floor, is integral with an external rocker that rotates in a hole of the mill wall (Figure 22). A cable is pulled by the rocker, and, through a pulley, closes the flume gate.

In the case of the Forno mill, a handwheel, positioned on the meal floor (Figure 23a), is integral with a shaft, keyed with a spur gear (Figure 23b). The driven gear is integral with a second shaft that pass through the wall of the mill, and is in turn keyed to a pinion in contact with a rack, whose translation regulates the vertical position of the flume gate (Figure 23a).

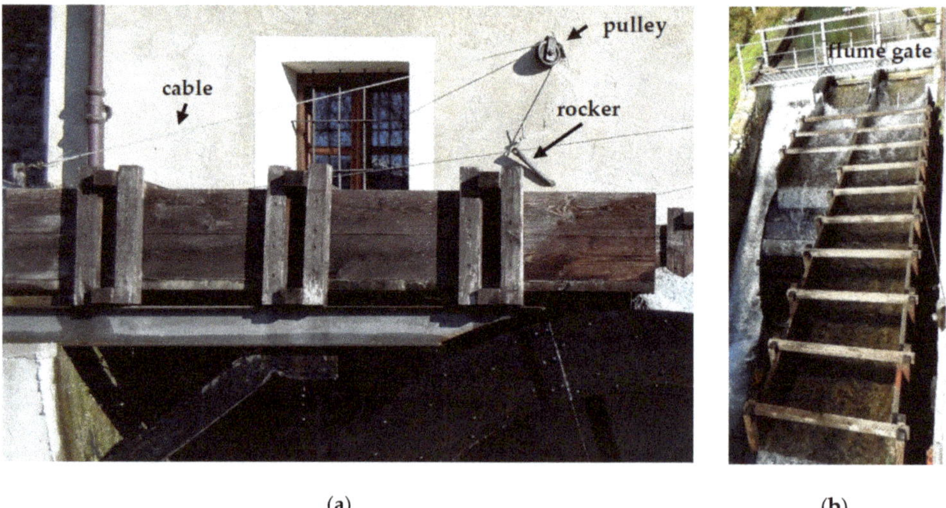

(**a**)   (**b**)

Figure 22. Regulating the supply flow rate in the Riviera mill: (**a**) The cable actuation; (**b**) the flume and the flume gate.

(**a**) (**b**)

Figure 23. Regulating the supply flow rate in the Forno mill: (**a**) The handwheel and the flume gate; (**b**) the gear reducer.

5. Conclusions

In the paper, the mechanical architecture of old watermills of Piedmont (Italy) is presented. Beyond the historical value of the work, the authors believe that the knowledge of the functional details of solutions used in the past allows us to fully understand their potential. They can also permit a full recovery of original tasks, as in the case of the Riviera mill, stimulate the design of new hybrid solutions, or inspire original and renewed uses. Old water wheels, for example, when for economic reasons are not suitable for direct actuation of the original machines, could sometimes be renovated for the production of electricity (micro hydropower generators). In fact, these machines present, in general, an interesting efficiency, between 70% and 85% (See Table 1). In particular, both the water wheels taken as example in this study have a good efficiency, close to 80%, and their restoration or reuse is therefore widely justifiable.

In all cases, for the sake of completeness, to the hydraulic study of the wheel a complete analysis of the mill must be added. In particular, it is necessary: (i) To go into the details of mechanical transmission, (ii) improve some mechanisms that are not very effective, (iii) evaluate the production potential, (iv) model or measure the torque and power characteristics of the whole transmission.

It is not always possible to accurately model the performance of the entire transmission. Water wheels, wooden gears, lubricated plain bearings have an intrinsically non-linear behavior, so that the uncertainty in determining the total efficiency of an old water mill can be significant.

In order to overcome this issue, it can therefore be useful, in perspective, to develop equipment aimed at the in situ experimental measurement of the mechanical characteristic of a water wheel and power transmission, and the definition of its overall efficiency. The portable device should consist of: (i) A unit for the connection with a mill power transmission; (ii) a braking unit capable of generating the load torque; (iii) a measurement unit, appropriately instrumented to determine the torque and power characteristic as a function of the rotation speed of the load shaft.

Author Contributions: Conceptualization, W.F., C.F. and R.R.; Methodology, W.F., C.F. and R.R.; Investigation, W.F.; Writing—original draft preparation, W.F.; Writing—review and editing, C.F. and R.R.

Funding: This research received no external funding.

Acknowledgments: Authors would like to thank the Cavanna family, owner of the Riviera mill, for their willingness, Davide Massucco for having carried out the survey of the Riviera mill and Paolo Cavagnero for some pictures of both mills.

Conflicts of Interest: The authors declare no conflict of interest.

References

1. Singer, C.; Holmyard, E.J.; Hall, A.R.; Williams, T.I. *A History of Technology*; Oxford University Press: New York, NY, USA, 1958.
2. Ferraresi, C.; Franco, W.; Quaglia, G. Human powered press for raw earth blocks. In Proceedings of the ASME 2011 International Mechanical Engineering Congress and Exposition, Denver, CO, USA, 11–17 November 2011.
3. Franco, W.; Quaglia, G.; Ferraresi, C. Experimentally based design of a manually operated baler for straw bale construction. In *Mechanisms and Machine Science*; Springer: Cham, Switzerland, 2017.
4. Olivero, R. *Macchine ad acqua. Mulini in Valle Maira [Water machines. Mills in the Maira Valley]*; I libri della Bussola: Dronero, Italy, 2009.
5. Rossi, C.; Russo, F.; Resta, F. Hydraulic Motors. In *History of Mechanism and Machine Science*; Springer: Cham, Switzerland, 2009.
6. Rossi, C.; Savino, S.; Timpone, F. An Analysis of the Hydraulic Saw of Hierapolis. In Proceedings of the 1st International Conference of IFToMM ITALY, Vicenza, Italy, 1–2 December 2016.
7. Yannopoulos, S.I.; Lyberatos, G.; Theodossiou, N.; Li, W.; Valipour, M.; Tamburrino, A.; Angelakis, A.N. Evolution of Water Lifting Devices (Pumps) over the Centuries Worldwide. *Water* **2015**, *7*, 5031–5060. [CrossRef]
8. Quaranta, E.; Revelli, R. Gravity water wheels as a micro hydropower energy source: A review based on historic data, design methods, efficiencies and modern optimizations. *Renew. Sustain. Energy Rev.* **2018**, *97*, 414–427. [CrossRef]
9. Rojas-Sola, J.I.; López-García, R. Computer-aided design in the recovery and analysis of industrial heritage: Application to a watermill. *Int. J. Eng. Edu.* **2007**, *23*, 192–198.
10. Rojas-Sola, J.I.; López-García, R. Engineering graphics and watermills: Ancient technology in Spain. *Renew. Energy* **2007**, *32*, 2019–2033. [CrossRef]
11. Rojas-Sola, J.I.; Domene-García, J. Engineering and computer-aided design: Study of watermills in southeastern Spain. *Interciencia* **2005**, *30*, 745–751.
12. Pujol, T.; Solà, J.; Montoro, L.; Pelegri, M. Hydraulic performance of an ancient Spanish watermill. *Renew. Energy* **2010**, *35*, 387–396. [CrossRef]
13. Pujol, T.; Montoro, L. High hydraulic performance in horizontal waterwheels. *Renew. Energy* **2010**, *35*, 2543–2551. [CrossRef]
14. Franco, W.; Ferraresi, C.; Revelli, R. Power Transmission and Mechanisms of an Old Water Mill. In *Mechanism and Machine Science*; Springer: Cham, Switzerland, 2019.
15. Chierici, P. *Fabbriche, Opifici, Testimonianze del lavoro. Storia e fonti materiali per un censimento in provincia di Cuneo [Factories, Workshops, work testimonials. History and material sources for a census in the province of Cuneo]*; Celid: Torino, Italy, 2004.
16. Müller, G.; Kauppert, K. Performance characteristics of water wheels. *J. Hydraul. Res.* **2004**, *42*, 451–460. [CrossRef]
17. Bach, C. *Die Wasserräder: Atlas (The water wheels: Technical drawings)*; Konrad Wittwer Verlag: Stuttgart, Germany, 1886. (In German)
18. Furnace, J. *Vintage Engraved Illustration*, Industrial Encyclopedia E.O. Lami. 1875.
19. Muller, W. *Die Eisernen Wasserrader, Erster Teil: Die Zellenrader und Zweiter Teil: Die Schaufelrader*; Veit & Comp.: Leipzig, Germany, 1899. (In German)
20. Quaranta, E.; Revelli, R. Output power and power losses estimation for an overshot water wheel. *Renew. Energy* **2015**, *83*, 979–987. [CrossRef]
21. Nuernbergk, D.M. *Overshot water wheels—Calculation and Design*; Verlag Moritz Schafer: Detmold, Germany, 2014.
22. Williams, A.; Bromley, P. New ideas for old technology-experiments with an overshot waterwheel and implications for the drive system. In Proceedings of the International Conference of Gearing, Transmission and Mechanical System, Nottingham, UK, 3–6 July 2000; pp. 781–790.

23. Quaranta, E.; Müller, G.; Butera, I.; Capecchi, L.; Franco, W. Preliminary investigation of an innovative power take off for low speed water wheels. In Proceedings of the International IAHR conference, New challenges in Hydraulic Research and Engineering, Trento, Italy, 12–14 June 2018.
24. Vidali, C.; Fontan, S.; Quaranta, E.; Cavagnero, P.; Revelli, R. Experimental and dimensional analysis of a breastshot water wheel. *J. Hydraul. Res.* **2016**, *54*, 473–479. [CrossRef]
25. Quaranta, E.; Revelli, R. CFD simulations to optimize the blades design of water wheels. *Drink. Water Eng. Sci.* **2017**, *10*, 27–32. [CrossRef]
26. Quaranta, E.; Revelli, R. Performance characteristics, power losses and mechanical power estimation for a breastshot water wheel. *Energy* **2015**, *87*, 315–325. [CrossRef]
27. Quaranta, E.; Revelli, R. Optimization of breastshot water wheels performance using different inflow configurations. *Renew. Energy* **2016**, *97*, 243–251. [CrossRef]

Article

Description and In-Vitro Test Results of a New Wearable/Portable Device for Extracorporeal Blood Ultrafiltration [†]

Paolo Boscariol [1], Giovanni Boschetti [1], Aldo Dalla Via [1], Nicola De Rossi [1], Mauro Neri [1,2], Ilaria Palomba [3], Dario Richiedei [1], Claudio Ronco [2,4] and Alberto Trevisani [1,*]

[1] Department of Management and Engineering, University of Padova, 36100 Vicenza, Italy; paolo.boscariol@unipd.it (P.B.); giovanni.boschetti@unipd.it (G.B.); aldo.dallavia@unipd.it (A.D.V.); nicola.derossi@unipd.it (N.D.R.); neri.mauro@gmail.com (M.N.); dario.richiedei@unipd.it (D.R.)

[2] Department of Nephrology, Dialysis and Transplantation and International Renal Research Institute of Vicenza (IIRIV), S. Bortolo Hospital, 36100 Vicenza, Italy; claudio.ronco@unipd.it

[3] Faculty of Science and Technology, Free University of Bozen/Bolzano, 39100 Bolzano, Italy; ilaria.palomba@unibz.it

[4] Department of Medicine (DIMED), University of Padova, 35128 Padova, Italy

[*] Correspondence: alberto.trevisani@unipd.it

[†] This paper is an extended version of our paper published in Boscariol, P.; Boschetti, G.; Dalla Via, A.; De Rossi, N.; Neri, M.; Richiedei, D.; Ronco, C.; Trevisani, A. RAP: A New Wearable/Portable Device for Extracorporeal Blood Ultrafiltration. In Proceedings of the Second International Conference of IFToMM Italy IFIT2018, Cassino, Italy, 29–30 November 2018.

Received: 15 April 2019; Accepted: 31 May 2019; Published: 4 June 2019

Abstract: This paper presents the design of Rene Artificiale Portatile (RAP), a novel wearable and portable device for extracorporeal blood ultrafiltration, capable of providing remote treatment of fluid overload in patients with kidney diseases and/or congestive heart failure. The development of the device is based on a new design paradigm, since the layout of the device is box-shaped, as to fit a backpack or a trolley case, differentiating it from other existing devices. The efficient layout and component placement guarantee minimalization and ergonomics, as well as an efficient and cost-effective use. The redundant control architecture of the device has been implemented to ensure a high level of safety and an effective implementation of the clinical treatment. The consistency of the design and its effective implementation are assessed by the results of the preliminary in-vitro tests presented and discussed in this work.

Keywords: wearable device; blood ultrafiltration; mechatronic device; renal replacement

---

1. Introduction

Fluid overload is a clinical condition in which the accumulation of water in the body cannot be excreted. One of the main consequences of fluid overload is electrolyte imbalance, in particular of sodium (whose normal level is diluted), which possibly leads to digestive problems, behavioral changes, brain damage, seizures and sometimes even coma [1]. In particular, the organs that lead to fluid overload are the kidneys and heart. Patients suffering from this pathological condition of fluid overload as the consequence of a renal disease or congestive heart failure need to be hospitalized and treated through dedicated extracorporeal blood purification therapies. In these treatments, patient's blood, in which water and toxins are accumulated in excess, is externally drawn by a specific vascular access, and through an extracorporeal circuit, filtered and purified by a specific medical device called hemodialyzer. Finally, the purified blood is re-infused into the patient. The term "blood purification therapy" refers to an array of different techniques that are applied according to the specific clinical status.

However, if the clinical aim is to remove excess water from patients, the specific technique is called ultrafiltration. Ultrafiltration aims at removing plasma water (the "ultrafiltrate", or "UF", in short) from blood through a semipermeable membrane, by means of a controlled pressure gradient [2]. In contrast to therapies such as hemodialysis, hemofiltration or hemodiafiltration, isolated ultrafiltration, with no use of replacement fluid, does not allow for a direct reduction of uremic toxins concentration.

The actuation and control of all processes to be performed during ultrafiltration therapy are generally carried out by devices designed for use in a hospital/clinic setting, although also some devices exist with an option for ultrafiltration at home. None of these devices, however, are small and lightweight enough to be carried around while performing the treatment. In order to extract blood from a patient's vascular circulation, a specific vascular access needs to be implanted. Except for the Artero-Venous fistula, the most used vascular access for this type of therapy is the dual lumen catheter [3], which allows taking out and re-infusing blood to the patient, puncturing the vessel only once. Typically, the catheter is positioned in the jugular or femoral vein, which are large-caliber blood vessels, in order to guarantee adequate blood flows flowing inside the extracorporeal circuit. Blood from the catheter circulates into the disposable extracorporeal circuit that shall be adapted with the interface of the so-called "hardware" of the ultrafiltration machine [2,4]. Since blood exits from a vein and returns to the same vein, an external pump generating a pressure gradient is necessary in order to perform extracorporeal circulation. The core of the filtering process takes place inside a disposable component called a hemodialyzer or hemofilter; it is a device constituted of a semipermeable membrane that allows the filtration of plasma water and suspended particulates (among which electrolytes like sodium and potassium) and the retention of components that should not be removed from patients, like proteins and red blood cells. The membrane is made of thousands (between 2500 and 10,000) hemo-compatible polymeric hollow fibers, porous on the surface, through which blood flows. In the space between fibers, the fluid removed (ultrafiltrate) is collected and then displaced. The physical phenomenon of filtration is called ultrafiltration, because pores have an average diameter smaller than 100 nm. The driving force promoting the separation is the hydrostatic pressure gradient across the membrane; a negative pressure is applied in the ultrafiltrate compartment of the filter through a pump (the ultrafiltration pump) in order to generate a positive transmembrane pressure [5].

Generally, ultrafiltration machines are bulky, and hence difficult to move. Furthermore, the management and set-up of these machines are not easy; consequently, a specialized and trained nursing and technical staff must setup and supervise each treatment procedure. Hence, ultrafiltration therapies generally require the hospitalization of patients [6]. The scientific community is joining the worldwide effort to overcome these issues by providing novel technological solutions for the extracorporeal ultrafiltration therapy. The design of portable and wearable devices for continuous ultrafiltration that can be operated without direct medical control, and/or even possibly monitored remotely, seems the most viable option. It has been extensively recognized [7–12] that the great potential of such portable/wearable medical devices is that they can implement a mild blood ultrafiltration therapy out of hospital over an extended stretch of time (such as over a full day), making this process more similar to physiological body water removing, reducing total therapy costs and improving patient quality of life.

Several prototypes with similar therapeutic targets have been developed to date [13–16]. Although many of them have contributed to scientific progress in this field, some technological limitations still restrict their clinical application and their industrialization [17–19]. Technological issues to be solved are mainly related to the requirements in combining safety for patients with miniaturization and low energy consumption. Hence, each new design must be developed with the greatest care to layout definition, to components choice, and to the implementation of a safe and robust control architecture.

A new mechatronic wearable device for extracorporeal blood ultrafiltration has been recently conceived by joint research project of theDepartment of Management and Engineering (DTG) of the University of Padova and the International Renal Research Institute of Vicenza (IRRIV) of the S. Bortolo Hospital. The device is named Rene Artificiale Portatile (RAP) after the Italian research program

"Rene Artificiale Portatile". Two papers describing the preliminary design of the RAP and of one of its more relevant components have been published recently [20,21]. This work provides a more in-depth description of the device, focusing on both the innovative aspects of its design which guarantee wearability, and on the control architecture. The results of preliminary in-vitro tests, which were performed to verify the safe and effective operation of the RAP device, are presented and discussed too.

2. Basic Operative Requirements

The RAP is intended to be a wearable/portable medical device that allows for safe out-of-hospital use under remote supervision. Consequently, the RAP should meet mandatory and strict requirements as formulated within international standards for safety and essential performance:

- IEC 60601-1 "Medical electrical equipment—Part 1: General requirements for basic safety and essential performance". This standard contains requirements for all medical electrical devices in general. These requirements pertain to safety aspects, human factor engineering and essential performance.
- IEC 60601-2-16 "Medical electrical equipment—Part 2-16: Particular requirements for basic safety and essential performance of haemodialysis, heamodiafiltration and heamofiltration equipment"
- IEC 60601-2-11 "Medical electrical equipment—Part 1-11: General requirements for basic safety and essential performance—Collateral Standard: Requirements for medical electrical equipment and medical electrical systems used in home care applications". This standard applies to specific aspects of the safety and essential performance of medical electrical equipment intended for use in home care applications, usually without continual professional supervision and sometimes temporarily used in the clinical environment.

A multi-disciplinary mechatronic design approach has been adopted: issues related to different domains (medical, mechanical, electrical and electronic domains), together with their coupling and mutual effects, have been considered.

First of all, the device must be powered autonomously for a sufficient lapse of time to allow for the usual mobility habits of patients treated. Since the RAP is intended to be used for continuous ultrafiltration therapy, 24 h a day, 7 days a week, adequate energy storage should be available, with a careful design of components and algorithms as well, in order to reduce energy consumption and heat generation. Furthermore, miniaturization, ergonomics, and low weight are obviously of paramount importance and have been taken into account during all the design phases.

Another key feature is the capability of providing an automatic regulation of the clinical parameters during the whole treatment according to the medical prescription. Hence, the device controller must manage the therapy procedure and log all the significant clinical data, which should be accessible in real-time for monitoring purposes in the hospital, where a dedicated wireless communication system can be setup. The data recorded during the whole treatment should then be collected after its conclusion and stored by the hospital IT department. The logged data include all the data that are relevant to the regulation of the process, and to the monitoring of the correct operation, and therefore include pressures in the extracorporeal circuit (access, pre-filter, return, ultrafiltrate, transmembrane pressure), the prescribed blood, ultrafiltrate and heparin flows, the ultrafiltrate volume, the air temperature inside the device housing, the battery level, and the list of warning and errors generated during the therapy.

It is essential that the applied vascular access avoids infections and formation of blood cloths; it is also necessary that connection and disconnection are easy to perform. Moreover, the extracorporeal tubes through which blood flows must be made of anti-thrombogenic materials and the total priming volume should be as low as possible. The material of membranes must be non-thrombogenic too, for minimizing the risk of coagulation and blood losses. The device must be equipped with a dedicated pumping system able to remove a volume of ultrafiltrate comparable to the one physiologically removed from the kidneys over the same amount of time. Therefore, the volume of liquid to be removed in 24-h is estimated between 1.5 and 2 L.

When blood encounters artificial materials, the coagulation system of blood is immediately activated, inducing a coagulation cascade. In order to reduce this phenomenon during extracorporeal blood circulation, a controlled and continuous infusion of anticoagulation drugs, such as heparin, must be administered into the circuit as well. Considering a heparin infusion flow ranging between 500 and 1500 U.I./h, and adopting a solution where heparin concentration is 500 U.I./mL, the desired infusion flow range of the heparin pump is between 1 and 2.5 mL/h.

The design procedure has been preceded by a detailed risk management analysis carried out according to the international standards ISO 14971 and IEC 60601-2-16, in order to define the safety devices and components that should be included in the RAP, thus allowing elimination of the non-strictly necessary ones. Some of the selected components that have been integrated in the system have been chosen among commercially available ones, some others have been specifically conceived, designed and implemented. Furthermore, a specific initial effort has been devoted to the distinction between reusable and disposable components. Clearly, all components that have contact with the patient's blood must be replaced after each use. This requirement has been addressed, as will be specified in Section 3, through a specific layout design.

The list of the strictly necessary components of the RAP includes:

- an air sensor to detect air bubbles in the blood line;
- a blood leak detector (BLD) to detect blood loss due to membrane rupture;
- three pressure sensors (access pressure, pre-filter pressure and return pressure sensors) to detect disconnections or blood coagulation;
- a pressure sensor in the ultrafiltrate line to detect coagulation in the filter;
- a fluid balance monitoring system to check the plasma water volume removed from the patient;
- a temperature sensor to monitor excessive temperature variations;
- a sensor detecting when the heparin reservoir is empty.

Furthermore, a clamp (safety valve) capable of automatically occluding the disposable circuit where blood flows is included in the design. This avoids any undesired infusion (e.g., clots, air) into the patient. It must be placed immediately before the blood return side of the catheter and must be operated whenever a safety hazard is detected. Since commercially available safety clamps are unsuitable for portable/wearable devices because they are large, heavy and high-power consuming, a customized and innovative design has been specifically applied for this component [21].

3. Layout of the RAP

The first and more evident design goal for the rap is keeping the overall size to a minimum [15]. This feature is dictated by the choice to improve the portability of other existing devices, by ensuring that the layout fits a backpack or a trolley case. Figure 1 shows a 3D - sketch of the virtual prototype, as well as a picture of the first working prototype of the device. The latter has been built using additive manufacturing technology. Some components are masked in the pictures since their patentability is currently under investigation.

The main feature of the design is the 'multiple layer' paradigm, which represents a significant departure from the more common 'single layer' paradigm used by most existing devices for ultrafiltration, such as the ones that are worn as a belt or as a sling [8]. This design introduces several advantages, both for patients (ease of carrying and use, both when standing or seating) and for healthcare professionals (simple installation of disposable components, easy setup of the clinical procedure, easy maintenance).

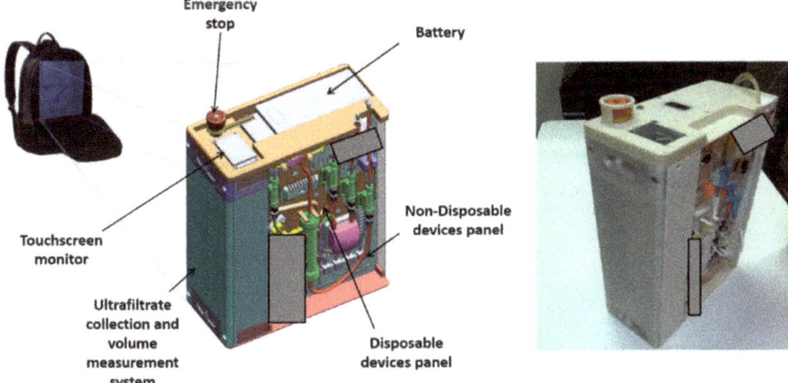

Figure 1. The Rene Artificiale Portatile (RAP) design: 3D sketch and manufactured prototype. Masked sections cover components where patentability is still under investigation at manuscript submission.

The mechanical layout is split into three layers, each one hosting a set of components. The first layer, which can be identified as the 'backbone' panel, hosts the main electronic components, including two microcontroller boards and a single-board microcomputer. The functionality of these devices, which implement the system control architecture, will be described in detail in Section 6. The CAD sketch and a picture of the first layer are shown in Figure 2.

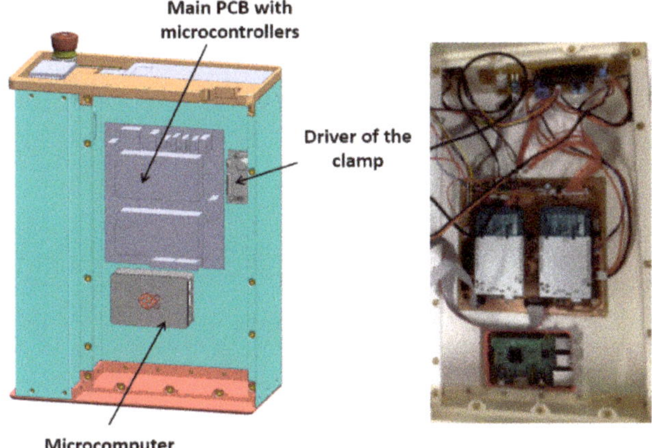

Figure 2. Internal panel with the main electronic components of the RAP: 3D sketch and manufactured prototype.

The 'intermediate' panel hosts other non-disposable components. The main ones are the peristaltic blood pump, the heparin infusion system, the custom-built electromechanical safety clamp, the air sensor, the temperature sensor and the BLD. These components are visible on the front of the panel, as shown in Figure 3. Looking at the backside of the panel, the ultrafiltrate pump driving circuit and the board for the signal conditioning circuit are visible.

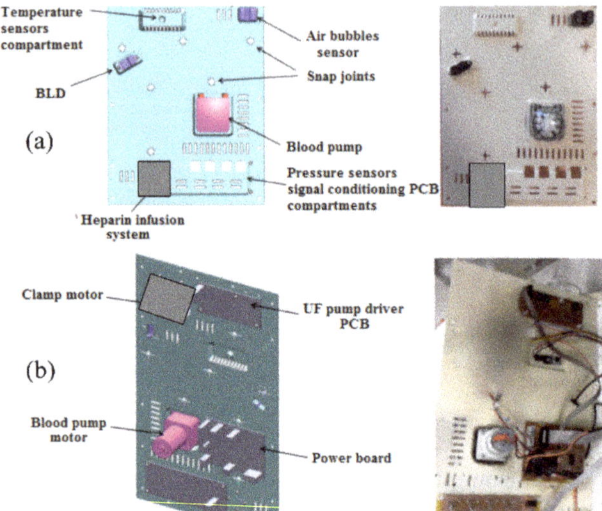

Figure 3. Front and back views of intermediate panel with the non-disposable components of the RAP: 3D sketch and manufactured prototype, (**a**) front side, (**b**) back side.

The third or external layer of the RAP hosts all the disposable components (the priming volume, included the filter, is almost 60 mL): the polyvinylchloride blood circuit, the hemofilter (priming volume 33 mL), the disposable elements of the heparin infusion system, the pressure sensors and the piezoelectric pump that displaces the ultrafiltration liquid.

The external layer, as shown in Figure 4, is designed as a 'replacement kit' or as a 'cartridge' system, that can be snapped in the device during the preparation of each treatment, and quickly disconnected and disposed after the use. The same figure also lists all the components of the blood line circuit.

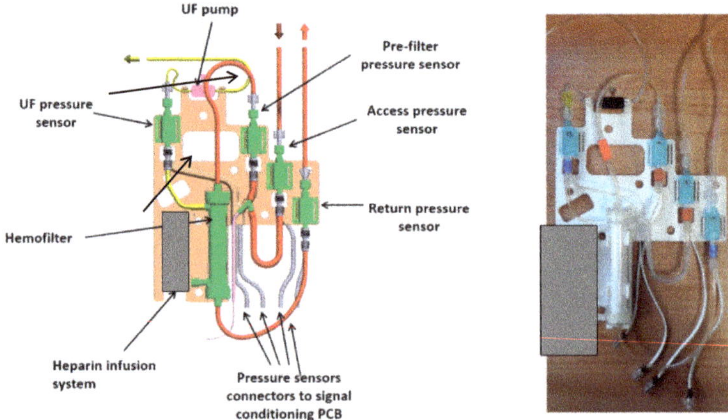

Figure 4. External panel with the disposable components of the RAP: 3D sketch and manufactured prototype.

The battery that powers the RAP is located in the cover panel (Figure 5), which hosts also a touchscreen display and an emergency stop button.

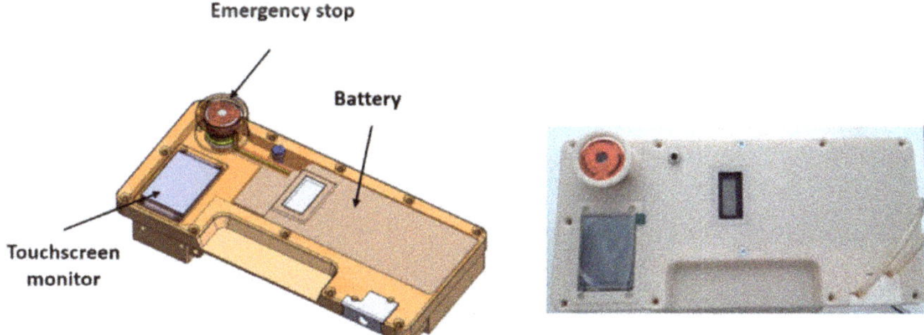

Figure 5. Covering panel on the top face of the RAP: 3D sketch and manufactured prototype.

The power source is a lithium battery with 75.5 Wh capacity, equipped with USB connection for quick charge technology. According to the results of several experimental trials, the average electric power consumption is equal to 13.9 W, and therefore the chosen battery can provide up to 5 h of running time. The design, however, gives the possibility of extending running time simply by including a small backup battery, allowing a "hot swap" main battery replacement without interruptions to the therapy, or by plugging the RAP into an AC outlet to recharge the battery.

The LCD touchscreen display mounted on the cover panel can be used for the treatment setup by healthcare professionals, as well as for monitoring of the main treatment parameters by patients. Patients can verify the correct operation of the device or the presence of a warning (e.g., by a warning message, such as 'empty the UF tank'), they can pause/resume the therapy, or they can follow the instructions prescribed by the therapeutic procedure. The LCD display, which is wired to the main board, can also be removed from its housing and placed on a shoulder strap for an easier use or monitoring.

The ultrafiltrate liquid is collected in a tank that is located on the left side of the device (see Figure 1). The tank, which can be easily removed to be emptied, has a capacity of 2.2 L and is equipped with a customized measurement system. Ultrafiltrate volume, which is measured by the RAP control architecture, is one of the therapy key parameters, and therefore must be monitored continuously during the whole clinic procedure.

The overall dimensions of the prototype are 405 by 300 by 140 mm, and the total weight is slightly below 5 kg with an empty tank, 7.2 kg with a full ultrafiltrate tank.

4. Control Architecture

The control architecture adopted for the RAP is shown in Figure 6.

The architecture has been carefully designed to meet both the restrictive requirements imposed by the application (the development of a miniaturized, energy efficient and wearable device which could be monitored remotely) and those imposed by the safety regulations of medical devices with the aim of minimizing risk for patients. The architecture can be split into two main sections, named the "program application section" and the "device control section", according to Figure 6. The main components of such an architecture are the microcomputer and the two microcontrollers boards, which work together to handle the RAP functionalities.

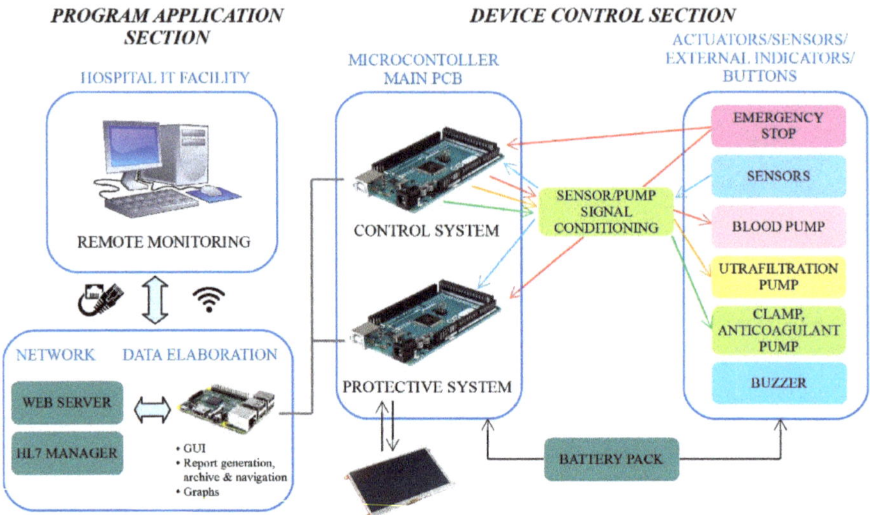

Figure 6. RAP control architecture.

The core of the program application section is run by the embedded Raspberry PI microcomputer, which runs the therapy management application, and implements the top layer of the software that runs on the RAP. The main tasks of the program application section are:

- management of communication with the device control section
- management of communication with an external PC through a web server and through the HL7 protocol
- logging of the data collected during the therapy
- management of treatment prescription and its transfer to the device control section
- management of the Graphical User Interface (GUI), accessible locally or through a wired/wireless LAN connection as a web page, that can be used for treatment set-up in hospital and for real-time monitoring of treatment data.

HL7 (Health Level 7, standard ISO 16527) is a specific standard protocol for exchanging information between medical applications. This standard defines a format for the transmission of health-related information, and therefore is commonly supported by the IT facilities of hospitals.

The device control section represents a lower layer of the control architecture that runs on the RAP, since it acts as an intermediate level between the 'high level' therapy management handled by the onboard microcomputer, and the 'bottom level' which includes the sensors, actuators, the LCD display and the power supply. The control section, however, does not simply act as a driver, since it takes care of running the actual control loop which ensures safe and correct operation of the RAP.

Each treatment is customized on the basis of the medical prescription through the GUI that runs on the Raspberry PI microcomputer, which translates them into a cluster of numeric set-points, which are in turn transferred to the device control section. Data sent from the device control section to the program application section include the key information on the status of the system (pressure values, amount of the ultrafiltrate removed, etc.). Such communication is performed through a USB connection between the microcomputer and the two microcontroller boards.

The redundancy of the control, which is a strict requirement imposed by the safety regulations, is implemented within the device control section. Each microcontroller independently performs the reading of all the sensors, and data are continuously compared to detect any possible mismatches between the two readings. The detection of a sensible discrepancy between readings, which suggests a measurement failure, triggers an emergency stop of the treatment. An emergency stop can also be triggered by the emergency button (also read by both microcontrollers), by the microcomputer, by the LCD, or by the device control section when a potential hazard is detected. It must be pointed out that the two microcontrollers run two different software codes, since not all operations are performed in a redundant fashion. Running two different codes also avoids the possibility of identical, and therefore undetectable, errors in both applications.

The two microcontroller boards are labelled as 'control system' and 'protective system' (see Figure 6) to highlight their different purposes. The 'control system' completely manages the ultrafiltration therapy, by reading the signals from the sensors and by providing the correct set points to the pumps. Sensor data are also transmitted to the microcomputer that stores them in a data log file, allowing complete monitoring of the therapy. The second microcontroller board, that is, the 'protective system', takes care of handling safety, mainly by performing the redundant reading of the signals from the sensors and performing the comparison with the data generated by the other board. In case of out of range measures or mismatches between the measures read by the two microcontrollers, a safety procedure is activated, imposing treatment interruption by stopping the pumps and closing the electro-mechanical safety clamp. At the same time, one bit of the communication array is set at a high logical level in order to communicate the current status immediately and to display an alarm message on the GUI. Furthermore, the protective microcontroller is connected to the LCD touchscreen, through which patients can immediately check the therapy status. The safety procedure can also be activated manually, by the emergency stop button placed on the top of the device and any alarms or warnings detected by the system are signaled through a buzzer.

In order to further improve the overall system safety, a watchdog hardware component has been fitted to continuously check the microcomputer status, ensuring that none of the microcontroller boards 'freeze'.

An accurate development of the control logic has been carried out as well; it has been developed based on clinical requirements, risk analysis, and after investigating thoroughly the behavior and the performances of some standard and reliable existing extracorporeal blood ultrafiltration machines.

The control logic has been structured in three main sequential operational phases to be performed. They are:

- Setup phase;
- Therapy phase;
- Termination phase.

All of these phases have been translated into software routines running on the microcontrollers and the microcomputer. Each phase has been further divided into relevant clinical procedures and technical steps to be executed. A graphic overview of the sequence of operations included in each phase is shown in Figure 7.

The setup phase consists of the following clinical procedures:

- switching on of the device: during this state, the system verifies the communication between the two microcontrollers and the microcomputer;
- mounting of the disposable components ('cartridge') on the non-disposable panel. Such a procedure needs to be performed in hospital by a trained healthcare professional;
- prescription of the therapy (remotely adjustable only by the physician);
- priming of the extracorporeal circuit: this is the most important procedure to be performed in the setup up phase. The automated priming is performed in order to fill the extracorporeal circuit with physiologic solution, remove all the air from the circuit, deposit some heparin along the inner surfaces of the hemofilter and tubes, and check if the circuit and the sensor have been connected properly;
- connection of the circuit to the patient catheter.

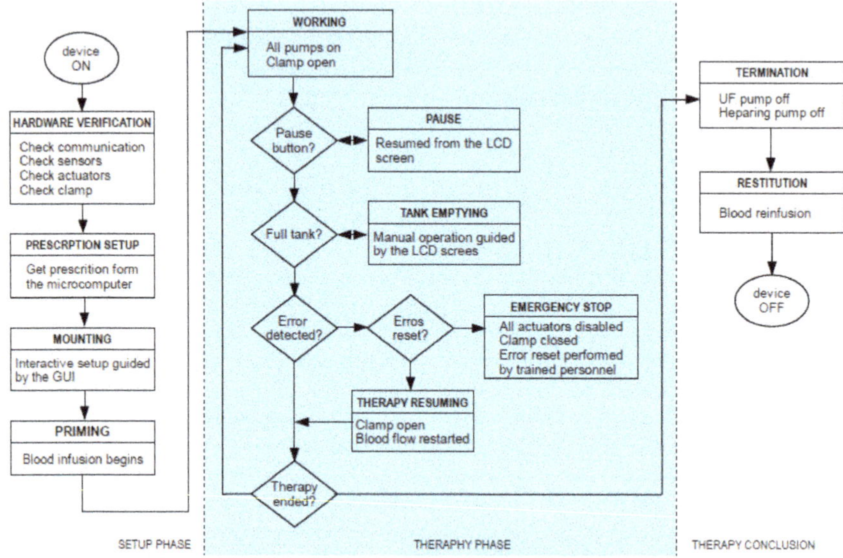

Figure 7. Workflow of the control architecture.

The setup phase is handled by the microcomputer, since it requires the input of several clinical parameters to correctly define the treatment according to the medical prescription. Such an operation is performed at the hospital by the healthcare assistants, who also take care of performing blood circuit priming, during which the conditions for starting extracorporeal blood circulation is ensured.

The second operational phase is the one corresponding to the ultrafiltration therapy. This phase is supervised by the program application section, that is, by the microcomputer, but theoretically the device control section could run the whole phase autonomously.

The first procedure to be performed in this phase is to initiate extracorporeal blood circulation, therefore starting the actual ultrafiltration treatment. If a hazardous situation is detected, the device automatically enters the state of alarm. In this state, the device automatically reacts by stopping the pumps and occluding the clamp based on the type of alarm and its risk. The alarm can be reset; if the problem is solved, the therapy can be carried on, otherwise the system re-enters the state of alarm.

The therapy phase can be paused at any time by stopping the blood pump. This can be done easily by touching a dedicated key on the touch sensitive LCD panel, or alternatively, through the GUI running on the microcontroller. The same methods also allow the therapy to resume. If the duration of the "pause" phase lasts more than 2 min, the machine raises an alarm for prolonged blood stagnation. The same check is made when the emergency stop button is pressed. Figure 8a shows examples of messages appearing on the LCD panel during a therapy; information fed back to patients includes the battery charge status and the amount of liquid accumulated in the ultrafiltration tank. The 'drain bag' button should be pushed just before manually emptying the tank in order to set up the device properly. The detection of a safety hazard triggers the immediate stop of the procedure, which is then signaled on the LCD screen, as can be seen in Figure 8b. Each possible safety hazard is identified by a numerical code, and each error event is logged.

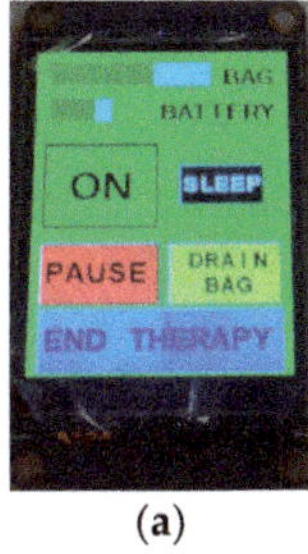

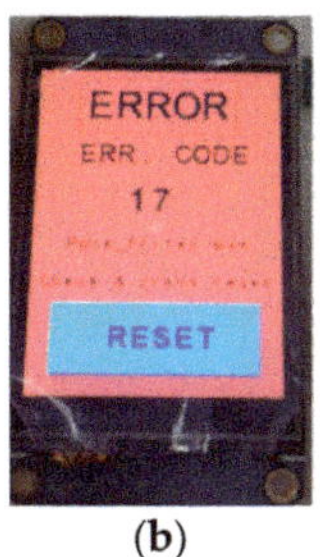

(a)   (b)

Figure 8. LCD screen during the therapy (**a**), LCD screen after a safety hazard detection (**b**).

The third and final phase of the therapy can, again, be triggered through either the LCD or the GUI; during this phase ultrafiltration ceases. It should be highlighted that this phase must be carried out in the hospital with a trained nurse who has to perform all the needed operations, in the proper order, as listed and guided by the GUI. At the beginning of the termination phase all pumps are stopped, thus ceasing ultrafiltration. After that, the blood along the external circulation line must be reinfused into the patient. To do this, the access line needs to be disconnected from the catheter and connected to a physiological solution; the RAP blood pump is then re-activated until all blood is removed from the circuit. The device can then be turned off, and the return tube can be disconnected from the catheter.

5. Preliminary Validation Tests

Preliminary in vitro tests have been carried out on the RAP to experimentally verify the effectiveness of the device components and of the device as a whole. A dedicate test bench platform was used for this purpose. The tests aimed to verify the following features of the RAP:

- reliability of pressure measurements in the RAP, by comparing the estimation with the measured values and with those of the test bench platform;
- reliability of the blood flow estimation procedure implemented in the RAP, by comparing the values measured by flow meters installed in the test bench platform;
- reliability of the temperature sensor used in the RAP, by comparing the measured values with those of the test bench platform;
- reliability of the ultrafiltrate measuring system used in the RAP, by comparing the estimated values with those obtained by an electronic scale;
- reliability of the power-bank battery used to supply the RAP for a duration of at least 5 h;
- mechanical resistance of the disposable tubes of the extracorporeal circuit for a duration of at least 5 h;
- biocompatibility, in terms of possible cytotoxicity induced by contact with all the disposable components of the circuit that will get into contact with blood.

All tests have been carried out by circulating 0.9% saline solution, with a density almost identical to water. The frequency of acquisition of pressure values measured by both devices has been equaled (20 Hz) in order to compare the two signals properly. Figure 9 shows the typical peristalsis-induced trend of the return pressure: the data collected by the test bench pressure measuring system are consistent with the ones measured by the RAP. This consistency has been checked, with similar results, for all three RAP pressure sensors (access, pre-filter and ultrafiltrate pressures). Based on these results, the choice of the pressure sensors and the signal conditioning procedure adopted in the RAP have been considered suitable.

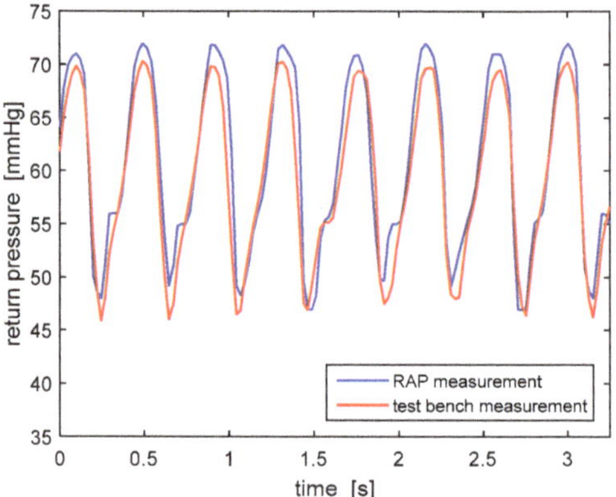

Figure 9. Return pressure values measured by RAP and test bench.

As a second step, the reliability of the blood flow estimation system used in the RAP, which is based on a real-time elaboration of the amplitude and frequency of the pre-filter pressure signal, has been verified. Figure 10 shows the comparison between the blood flow rate estimated according to the algorithm run by the RAP and the one measured by the test bench for three different set point values (60, 80 and 100 mL/min). The flow rate estimated by the RAP (in red) slightly differs from the one measured by the flow meter. In particular, the worst-case estimation error is equal to 9.12%, which is well below the accuracy limits highlighted by the risk analysis process.

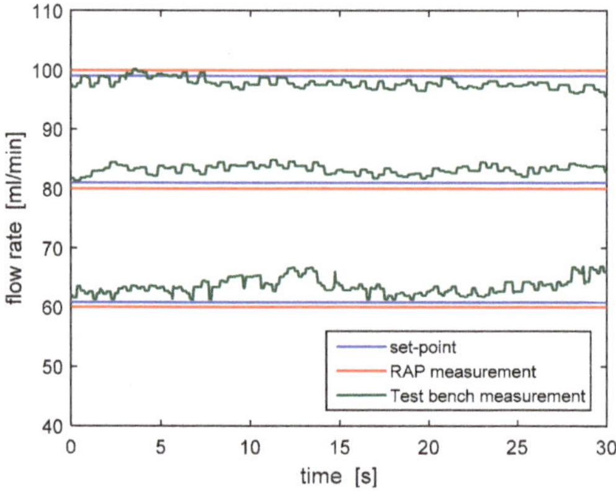

Figure 10. Blood flow rate estimated in the RAP and measured by the test bench at different set values (60, 80 and 100 mL/min).

Subsequently, the temperature sensor of the RAP, which measures the temperature of the air inside the device housing, was validated. Temperature values have been recorded every 10 min for a total treatment duration of 5 h and 12 min. Figure 11 shows the comparison of temperature values read by the two measurement devices, with a room temperature equal to 25.5 °C. The maximum percentage error recorded was 5%; according to the results of the risk analysis this measurement error does not provide any actual safety hazard. The temperature measurements show that the initial temperature, 25 °C, increased to the stable measure of 30 °C within the first hour of operation, thus no overheating was detected.

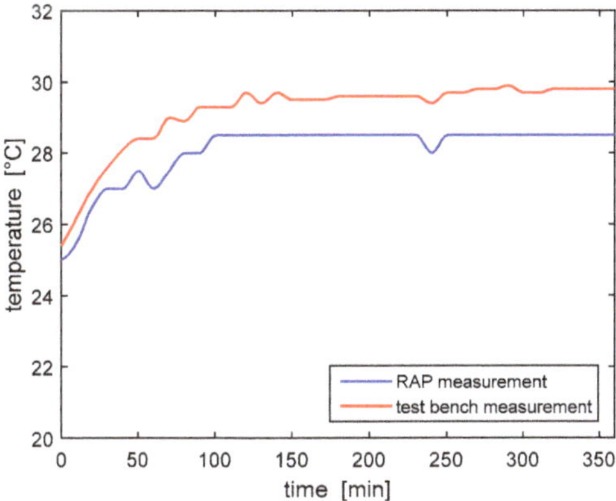

Figure 11. Comparison of temperature values read by the RAP and the test bench.

After more than 5 h of continuous operation, no safety hazards were detected, providing a first proof of the reliability of the device. During the same test, the estimation of the ultrafiltration volume has been checked against the reading of an electronic scale. The ultrafiltrate estimation system performs a state estimation by comparing data generated by ultrasonic sensors and accelerometers. No further details are provided here, for patentability issues.

Figure 12 shows the measurement collected during the 5-h trial; the plot compares the UF volume set point (equal to 5 mL/min), to the estimated and measured volumes. With the exception of a short initial interval, when it was not possible to estimate the extracted volume due to the design of the UF tank, the liquid removal rate was estimated correctly and follows the desired one with a minimum error. This test therefore validated both the reading and the regulation of the ultrafiltrate removal rate.

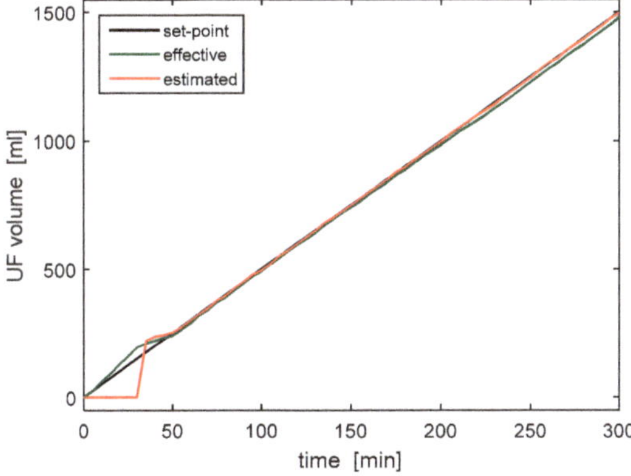

Figure 12. Ultrafiltrate volumes set in the device (black), estimated by RAP (red) and effectively measured by external scale (green).

Finally, the possible induction of cytotoxicity has been evaluated, to verify that the device does not lead to excessive deterioration of the blood tissues. The test was carried out by circulating 500 mL of cell medium for 6 h: it is a special-purpose fluid with density and viscosity very similar to water and containing many products necessary for cell growth and maintenance of the physical-chemical conditions such as pH and osmolarity. Cell viability was assessed by measuring the number of viable cells and determined by trypan blue (Sigma, St. Louis, Mo., USA). Fluid samples were collected at the beginning of the test and then after 30, 60, 180 and 360 min.

A specific cell line, monocyte line U937, which is considered to be the gold standard for cytotoxicity analyses, was incubated for 24 h in the samples taken. At the end of the 24 h, via a cytofluorometer, values of viability, necrosis (cell death due to artificial causes) and apoptosis (cell death due to natural causes) of the incubated monocytes were evaluated. These values were compared with a control sample. The analysis of the samples was translated into percentage of viability, necrosis and apoptosis, as displayed in Figures 13–15. Data show, for all three parameters, similar values with respect to the control samples within 6-h running time of the procedure, highlighting minimal cytotoxicity. These results confirm the absence of adverse reactions, potentially due to high mechanical stress, increased temperature, release of contaminants or reaction products as a consequence of the contact with external materials and components of the RAP prototype.

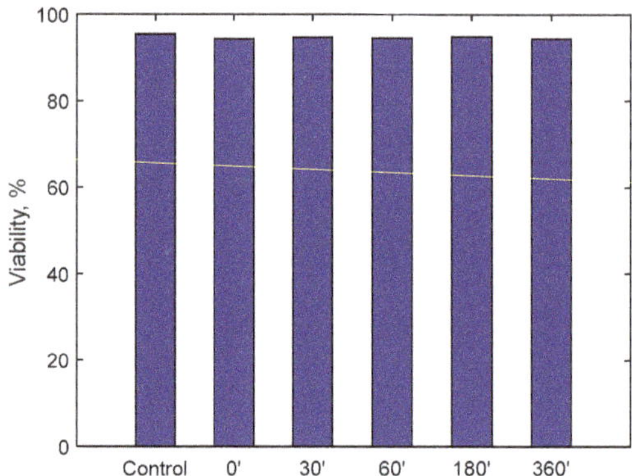

Figure 13. Viability on monocytes cell line U937 after incubation with fluid circulated in the RAP for up to 360 min.

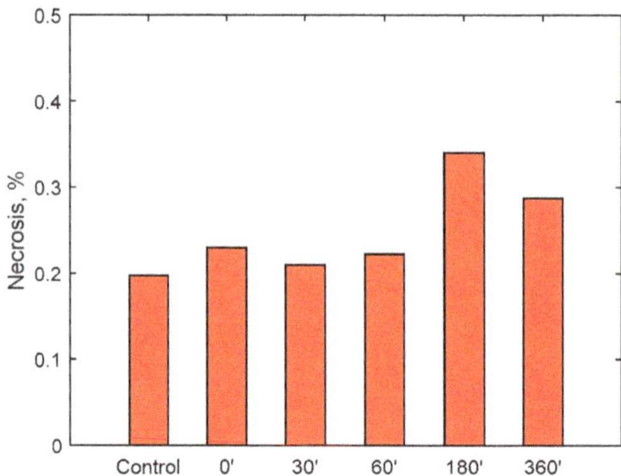

Figure 14. Necrosis on monocytes cell line U937 after incubation with fluid circulated in the RAP for up to 360 min.

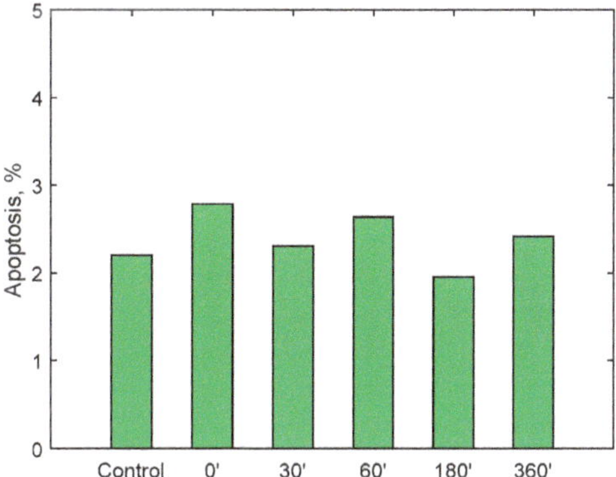

Figure 15. Apoptosis on monocytes cell line U937 after incubation with fluid circulated in the RAP for up to 360 min.

6. Conclusions

This work has introduced the first pre-clinical prototype of a novel wearable/portable device for blood ultrafiltration, named RAP. The design mixes additive manufacturing technologies, off-the-shelves components (pumps, actuators, membranes, sensors, electronic devices), and some custom design components, such as a novel electromechanical clamp and a sensorized ultrafiltrate tank. Another significant novelty of the RAP device lies in its layout paradigm, based on three main layers, arranged to fulfill the space occupation specifications defined at the earliest stages of the development. The final design achieved the best tradeoff between miniaturization and ergonomics through a box-shaped design that easily fits a trolley case or a backpack.

The box-like shape is the consequence of stacking three layers of components, one for the disposable devices, one for the non-disposable components and one for the main electronic boards. This arrangement significantly simplifies the maintenance of the device, as well as the procedures that precede and follow each treatment run, achieving the minimization of the time required to complete the operations performed by the trained personnel, which should take care of the initial and final phases of the therapy.

The control architecture has been outlined in detail, highlighting the hardware and software solutions chosen to ensure safe operation of the device.

The functionality of the device has been tested by specific in-vitro tests, through which the correct operation of the critical elements of the design was assessed. The correct and reliable operation of the RAP is complemented by the low rate of hemolysis and the negligible levels of cytotoxicity displayed by the specific test reported here.

The device is now ready for further in-vitro testing, and for in-vivo testing. The device introduced here is the first working prototype, conceived to validate a new design; there is room for miniaturization by deepening integration though a custom electronic design and through the new design of disposable components, such as filters and pressure sensors.

Author Contributions: Conceptualization, M.N., C.R. and A.T.; Data curation, P.B., M.N. and I.P.; Formal analysis, P.B., M.N., D.R. and A.T.; Funding acquisition, C.R. and A.T.; Investigation, P.B., G.B., A.D.V., N.D.R., M.N., I.P., D.R., C.R. and A.T.; Methodology, M.N. and A.T.; Project administration, C.R. and A.T.; Resources, N.D.R., M.N., C.R. and A.T.; Software, P.B., G.B., A.D.V. and M.N.; Supervision, C.R. and A.T.; Validation, P.B., G.B., A.D.V., N.D.R., M.N., I.P., D.R., C.R. and A.T.; Writing—original draft, P.B. and M.N.; Writing—review & editing, P.B., M.N. and A.T.

Funding: The authors acknowledge the financial support by Fondazione Cariverona through the research grant "RAP" 2014-2017, ref. 2014.0850.

Conflicts of Interest: The authors declare no conflict of interest.

References

1. Maxine, A.P.; McPhee, S.J.; Rabow, M.W. *Current Medical Diagnosis & Treatment*; McGraw-Hill Medical: New York, NY, USA, 2011; p. 51.
2. Neri, M.; Alliance, O.B.O.T.N.S.I. (Nsi); Villa, G.; Garzotto, F.; Bagshaw, S.; Bellomo, R.; Cerdá, J.; Ferrari, F.; Guggia, S.; Joannidis, M.; et al. Nomenclature for renal replacement therapy in acute kidney injury: Basic principles. *Crit. Care* **2016**, *20*, 318. [CrossRef] [PubMed]
3. Dahle, T.G.; Blake, D.; Ali, S.S.; Olinger, C.C.; Bunte, M.C.; Boyle, A.J. Large Volume Ultrafiltration for Acute Decompensated Heart Failure Using Standard Peripheral Intravenous Catheters. *J. Card Fail.* **2006**, *12*, 349–352. [CrossRef] [PubMed]
4. Villa, G.; Neri, M.; Bellomo, R.; Cerda, J.; De Gaudio, A.R.; De Rosa, S.; Garzotto, F.; Honore, P.M.; Kellum, J.; Lorenzin, A. Nomenclature for renal replacement therapy and blood purification techniques in critically ill patients: Practical applications. *Crit. Care* **2016**, *20*, 283. [CrossRef] [PubMed]
5. Ismail, A.F.; Abidin, M.N.Z.; Mansur, S.; Zailani, M.Z.; Said, N.; Raharjo, Y.; Rosid, Y.R.; Othman, M.H.D.; Goh, P.S.; Hasbullah, H. Hemodialysis membrane for blood purification process. In *Membrane Separation Principles and Applications*; Rahman, I., Matsuura, O., Eds.; Elsevier: Amsterdam, The Netherlands, 2019; pp. 283–314.
6. Nexight Group. Technology Roadmap for Innovative Approaches to Renal Replacement Therapy. Available online: https://www.asn-online.org/membership/BlastEmails/files/KHI_RRT_Roadmap1.0_FINAL_102318_web.pdf (accessed on 31 May 2019).
7. Topfer, L.A. Wearable artificial kidneys for end-stage kidney disease. In *CADTH Issues in Emerging Health Technologies*; Canadian Agency for Drugs and Technologies in Health: Ottawa, ON, Canada, 2016.
8. Salani, M.; Roy, S.; Fissell, W.H. Innovations in Wearable and Implantable Artificial Kidneys. *Am. J. Kidney Dis.* **2018**, *72*, 745–751. [CrossRef] [PubMed]
9. Kooman, J.P.; A Joles, J.; Gerritsen, K.G. Creating a wearable artificial kidney: Where are we now? *Expert. Rev. Med. Devices* **2015**, *12*, 373–376. [CrossRef] [PubMed]
10. Beal, M.F.; Kowall, N.W.; Swartz, K.J.; Ferrante, R.J.; Martin, J.B. Differential sparing of somatostatin-neuropeptide y and cholinergic neurons following striatal excitotoxin lesions. *Synapse* **1989**, *3*, 38–47. [CrossRef] [PubMed]
11. Leonard, E.F. Technical approaches toward ambulatory ESRD therapy. *Semin. Dial.* **2009**, *22*. [CrossRef] [PubMed]
12. Leonard, E.F.; Cortell, S.; Jones, J. The Path to Wearable Ultrafiltration and Dialysis Devices. *Blood Purif.* **2011**, *31*, 92–95. [CrossRef] [PubMed]
13. Gura, V.; Rivara, M.B.; Bieber, S.; Munshi, R.; Smith, N.C.; Linke, L.; Kundzins, J.; Beizai, M.; Ezon, C.; Kessler, L.; et al. A wearable artificial kidney for patients with end-stage renal disease. *JCI Insight* **2016**, *1*. [CrossRef] [PubMed]
14. Bazaev, N.A.; Dorofeeva, N.I.; Zhilo, N.M.; Streltsov, E.V. In vitro trials of a wearable artificial kidney (WAK). *Int. J. Artif. Organs* **2017**, *41*, 84–88. [CrossRef] [PubMed]
15. Humes, H.D.; Buffington, D.; Westover, A.J.; Roy, S.; Fissell, W.H. The bioartificial kidney: Current status and future promise. *Pediatr. Nephrol.* **2014**, *29*, 343–351. [CrossRef] [PubMed]
16. Ronco, C.; Fecondini, L. The Vicenza wearable artificial kidney for peritoneal dialysis (ViWAK PD). *Blood Purif.* **2007**, *25*, 383–388. [CrossRef] [PubMed]
17. Lee, D.B.; Roberts, M. A peritoneal-based automated wearable artificial kidney. *Clin. Exp. Nephrol.* **2008**, *12*, 171–180. [CrossRef] [PubMed]
18. Ronco, C.; Davenport, A.; Gura, V. The future of the artificial kidney: Moving towards wearable and miniaturized devices. *Nefrología* **2011**, *31*, 9–16. [PubMed]
19. Kim, J.C.; Garzotto, F.; Nalesso, F.; Cruz, D.; Kim, J.H.; Kang, E.; Kim, H.C.; Ronco, C. A wearable artificial kidney: Technical requirements and potential solutions. *Expert Rev. Med. Devices* **2011**, *8*, 567–579. [CrossRef] [PubMed]

20. Boschetti, G.; Dalla Via, A.; De Rossi, N.; Garzotto, F.; Neri, M.; Pamato, L.; Ronco, C.; Trevisani, A. Conceptual Design of a Mechatronic Biomedical Wearable Device for Blood Ultrafiltration. Mechanism and Machine Science. *Adv. Ital. Mech. Sci.* **2017**, *47*, 89–96.

21. Boscariol, P.; Boschetti, G.; Caracciolo, R.; Neri, M.; Richiedei, D.; Ronco, C.; Trevisani, A. Design of a Miniaturized Safety Clamping Device for Portable Kidney Replacement Systems. *Adv. Ital. Mech. Sci.* **2017**, *47*, 79–87.

 *machines* 

Article

Flexible Fingers Based on Shape Memory Alloy Actuated Modules †

Daniela Maffiodo * and Terenziano Raparelli

Department of Mechanical and Aerospace Engineering, Politecnico di Torino, 10129 Torino, Italy; terenziano.raparelli@polito.it
* Correspondence: daniela.maffiodo@polito.it; Tel.: +39-0110905984
† This paper is an extended version of our paper Maffiodo, D., Raparelli, T.; Comparison among different modular SMA actuated flexible fingers In Proceedings of the The 2nd IFToMM ITALY Conference, Cassino, Italy, 29–30 November 2018.

Received: 9 May 2019; Accepted: 7 June 2019; Published: 10 June 2019

Abstract: To meet the needs of present-day robotics, a family of gripping flexible fingers has been designed. Each of them consists of a number of independent and flexible modules that can be assembled in different configurations. Each module consists of a body with a flexible central rod and three longitudinally positioned shape memory alloy (SMA) wires. When heated by the Joule effect, one to two SMA wires shorten, allowing the module to bend. The return to undeformed conditions is achieved in calm air and is guaranteed by the elastic bias force exerted by the central rod. This article presents the basic concept of the module and a simple mathematical model for the design of the device. Experimental tests were carried out on three prototypes with bodies made of different materials. The results of these tests confirm the need to reduce the antagonistic action of the inactive SMA wires and led to the realization of a fourth prototype equipped with an additional SMA wire-driven locking/unlocking device for these wires. The preliminary results of this last prototype are encouraging.

Keywords: shape memory alloy; SMA wires; flexible actuator; modular actuator; mathematical model; experimental test

---

1. Introduction

Shape memory alloys (SMAs) are a particular class of metal alloys characterized by two properties: (1) the shape memory effect (SME), the ability to recover a preset geometric shape when subjected to an appropriate temperature change; and (2) superelasticity (SE), the ability to withstand large deformations (up to 10–15% compared to the initial configuration) without producing permanent effects within a certain range of temperatures. The first discovery of these phenomena dates back to 1932 thanks to the studies of Chang and Read on the AuCd alloy, and in 1938 the transformation was studied in brass (CuZn). However, it was only in 1962 that Buehler discovered the SME in NiTi alloy, and it was from then that actual research on its metallurgy and on its practical applications began. Many other alloys that presented these properties were analyzed, but from the point of view of applications, the most interesting and useful proved to be those of the NiTi group and the Cu alloys [1].

Since then, SMAs have been used in many fields of engineering: robotics [2–7], biomedical engineering [8–10], and structural engineering [11,12]. The field of innovative robotics seems to be particularly suited to exploiting the advantages that these materials can provide. These advantages are a high power/weight ratio [13], sensing ability, remotability, low driving voltage, simplicity, cleanliness, and silent actuation. Some of these advantages are emphasized when the device decreases in size. On the other hand, they also present disadvantages, such as low energy efficiency, fatigue problems, history-dependent characteristics, and low actuation frequency [14]. Much research attempted to overcome these disadvantages, in particular, by implementing different controls in order to obtain a stable and repeatable behavior of their devices [15–17].

Today, industrial automation is entering a new phase, that of the Internet of things (IoT) and the smart factory. Robotics, which is one of the ingredients of this revolution, has the increasing need to find non-conventional solutions. Since the aim is to allow humans and machines to work together without protective barriers, manipulators must also be designed for cooperation.

The end-effector mounted on a manipulator wrist can be a working tool or a gripper. The traditional parallel or angular grippers, with two elements facing each other and operating a parallel or angular movement, are widely used in robotics since they are simple and generally economical [18,19].

Dexterous hands usually have an anthropomorphic shape and the object is grasped by closing mechanical fingers. They are more versatile and are preferred for gripping non-symmetrical objects [20–23]. The actuation system of the grippers can be traditional (electrical, pneumatic) or innovative, like piezo-actuation [24] or SMA actuator-based [16,17,25].

Moreover, flexible actuators are an interesting solution to use as fingers in the gripping hands, in particular for cooperative robots. Different solutions, operated using electric current, hydraulic fluid pressure, or pneumatic pressure [26–28], have been developed, but SMA-operated flexible fingers can also be found in the literature. In general, the solutions implemented with electric motors are less light, smooth, and silent compared to solutions with shape memory materials. Regarding the solutions with hydraulic and pneumatic actuation, they can be used only in environments where the corresponding generation plant is present. Devices with SMA actuation therefore appear to be promising where lightness both for energy system demand and the device itself is required.

As an example, Yang et al. [29] created a flexible device made of three SMA springs embedded off-axially and movably in a silicone rubber rod. A large deformation of the bending actuators are obtained. Drawbacks of their solution are the cooling time, increased by the embedded solution, and friction between the silicone and spring during the motion. Torres-Jara et al. [30] developed a compliant modular actuator based on different arrays of the same simple unit based on a folded sheet of SMA. Linear, rotational and surface actuators are obtained, depending on a different assembly. This is a very interesting solution, both for the design of the object and its performance and for the idea of amplifying the work space by means of the series or parallel composition of a certain number of modules.

The idea of a flexible actuator and that of a modular actuator have been merged together to create the new family of flexible modular actuators presented in this paper. Each module is small, light, and can be variously assembled to allow for the creation of fingers having different shapes and characteristics to satisfy a wide range of needs. Depending on the task, it will be possible to assemble a different number of fingers to constitute various grippers, from the simplest with two fingers for the grasping of simple objects to the anthropomorphic five-finger hand.

This study aims to contribute to this growing area of research by proposing a mathematical model for the design of a flexible actuator, the design and experimental tests performed on four actuators, and the comparison of their different behaviors.

2. SMA-Actuated Module

A modular actuator based on shape memory wires is sketched in Figure 1a. The module has a length of 40 mm and is composed of a central rod (1) with a lower base (2), an upper base (3), and two intermediate disks (4). Three Nitinol SMA wires (5) (diameter 250 μm) are longitudinally placed. One end of each wire is fixed to the lower base and passes through holes in the intermediate disks running parallel to the central rod. This end is now looped through a hole on the upper base and then returns to the lower base where it is fixed. Suitable screws (not shown) placed on the upper base allow the proper tensioning of the SMA wires. Each of the three wires is positioned at 120° from the others in order to allow the module to bend in any direction when one or more wires are actuated. The module bends when the wire is heated, e.g., by means of the Joule effect, which causes the shortening of the wire. When cooled and applying a bias force, the wire is stretched to the original shape and the module comes back to the undeformed shape. This bias force is exerted in this case both by the central rod and by the inactive wires. The wires are not embedded in the structure so as to obtain faster cooling.

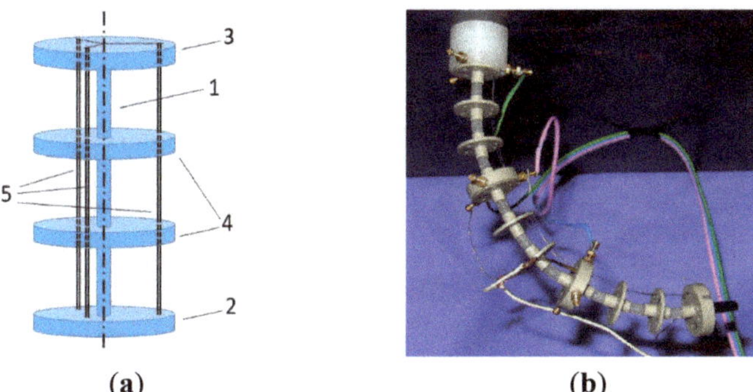

(a) **(b)**

Figure 1. (**a**) Sketch of one module: 1) central rod, 2) lower base, 3) upper base, 4) intermediate disks, and 5) three SMA wires @120°; (**b**) Example of a finger composed by three modules in series.

The modules can be joined in order to assemble fingers. This requires a rigid support that constitutes the base on which to arrange the electric power cables, switches, and drives for the control of the structure. As an example, Figure 1b shows a finger made of three modules. The heating of one wire will bend the module along one of three *main directions* (directions 1, 2, and 3 in Figure 2). The heating of two SMA wires will bend the module along three *secondary directions* (directions 1-2, 2-3, and 1-3 in Figure 2).

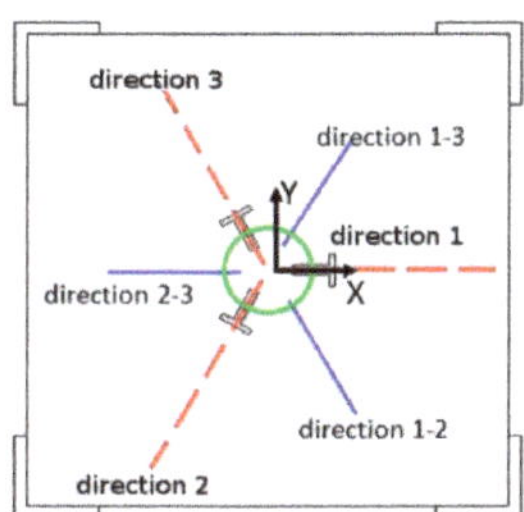

Figure 2. View from above of a module with actuator displacement directions.

The main advantage of this arrangement is that it allows the actuator to move within a volume whose projection on the plane perpendicular to the actuator is 360°; that is, it is possible to carry out all the desired positioning on this plane within the work space of the device.

However, the presence of the other inactive SMA wires, which by their function will be called *antagonists*, constitutes a physical limit to deformation. Solutions to this limitation are investigated in Section 6.

Figure 3 shows a rendered image of a three-finger gripper and its possible operation.

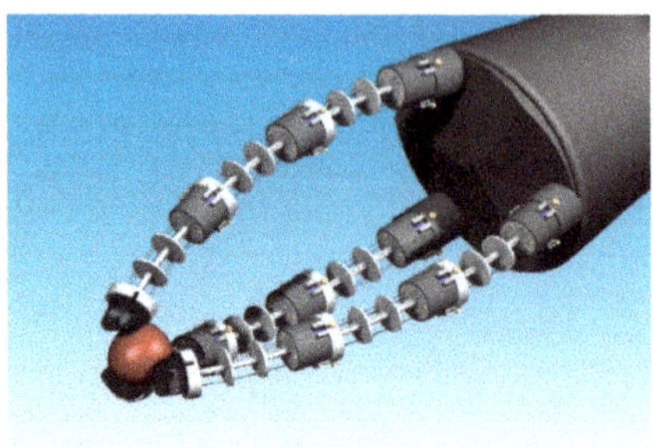

Figure 3. Three-finger gripper.

3. Mathematical Model

To evaluate the behavior of a three-module finger, and in particular, the contribution of the antagonistic wires, a numerical analysis was performed. This analysis involves the construction and development of a mathematical model of the system and its implementation and solution. A simplified mathematical model, in which the contribution of the antagonistic wires was neglected, is presented in Maffiodo and Raparelli [31].

This model consists of two parts to distinguish the different behavior in the case of a single SMA wire actuation (motion along the main directions, see Figure 2) and the actuation of two wires (movements along the secondary directions, see Figure 2) where the forces involved and the position of the latter with respect to the center of the device axis varies. Figure 4 shows the sketch of the forces acting on a module in these two different situations.

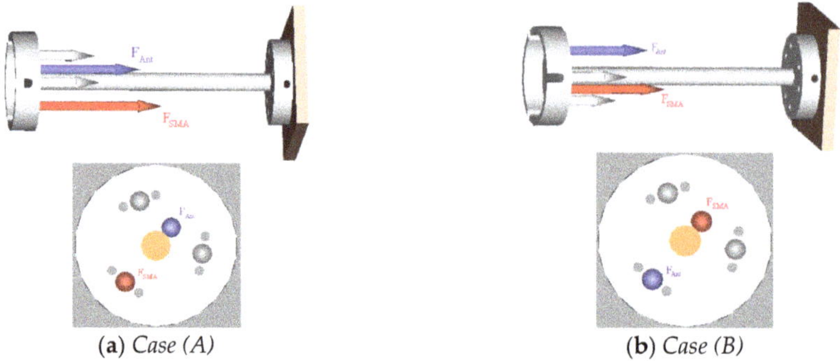

| (a) *Case (A)* | (b) *Case (B)* |

Figure 4. Sketch of the forces acting on a module: (**a**) lateral view and plant view in the case of one-wire activation (A), (**b**) lateral view and plant view in the case of two-wire activation (B).

Case A

The displacement along the main directions, as mentioned, is achieved by operating the SMA wires one at a time. In this case, therefore, there will be a force given by the active SMA wire and a double antagonist action determined by the two SMA wires at rest, which is added to the elastic antagonist action determined by the central rod (Figure 4a).

To build a simplified 2D model, the following assumptions were made. With reference to the sketch of Figure 4a, the force of action of the SMA wires (shown in red) is applied to the midpoint of the segment that connects the centers of the two holes in which the SMA wire is running and is directed along the wire itself in the shown direction. The two antagonist forces (shown in gray) are also applied in the midpoints of the segments that connect the centers of the two holes in which the two antagonist SMA wires are running, and are directed along the direction of the wires themselves and have the sense represented in the picture. These forces remain parallel to each other during the module deformation; therefore, instead of considering their separate effect, they are replaced by an overall antagonistic force (represented in blue). This antagonistic force has direction and sense equal to the direction of the two antagonistic forces, it has a magnitude that is the sum of the two, and its point of application is located at the midpoint of the segment that connects the two points of application of the two original antagonistic forces. As can be seen in Figure 4a, the actuating force exerted by the SMA wire, the overall antagonist force, and the axis of the actuator are then in the same plane, and therefore the deformed actuator will move along this plane.

Case B

The movement along the secondary directions is obtained by operating the SMA wires two at a time. In this case, therefore, there will be a force due to the two active SMA wires and a single antagonistic action determined by the inactive SMA wire, which is added to the elastic antagonist action determined by the central rod (Figure 4b). To build the 2D model, with reference to the sketch of Figure 4b, each one of the two SMA forces (shown in gray) is applied at the midpoint of the segment that joins the centers of the two holes along which the SMA wire runs. Similarly to the previous case, these forces remain parallel to each other throughout the deformation; therefore, instead of considering their separate effect, they are replaced by a total SMA force (represented in red). This force has direction and sense equal to the direction of the two SMA forces, it has a magnitude that is the sum of the two, and its point of application is located at the midpoint of the segment that connects the two points of application of the two SMA forces. The antagonist force (represented in blue) is generated by a single wire (the wire at rest); it is applied at the midpoint of the segment that connects the center of the two holes in which the antagonist SMA wire runs, as shown in Figure 4b.

Also in this case, the total SMA force, the antagonistic force, and the actuator axis are in the same plane; therefore the deformed actuator will move along this plane.

With these assumptions, in both cases the actuator is subjected to forces lying on a plane that will be called the "deformation plane." Therefore, the model is a two-dimensional (2D) model, based on the analysis of the actions during the actuation in the two cases, and it is subsequently extended to the other directions thanks to the symmetry that the actuator presents.

Moreover, in the model, other simplifying hypotheses have been made. The first one concerns the upper head of the actuator, which even after deformation, is supposed to remain perpendicular to the central rod; the second one consists in neglecting the inertia of the system.

Figure 5 is a sketch of the 2D model.

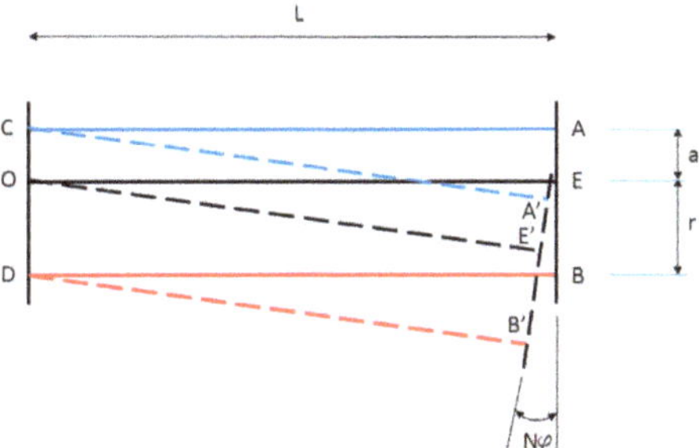

Figure 5. Sketch of the 2D model.

The axis of the actuator is represented by the segment OE, having length L. A generic deformed condition brings the point E (end-effector) to the position E′.

To study this deformation, the actuator is modelled as a set of N rigid bodies having a length equal to L/N each. These elements are connected to each other via torsional springs [31]. For simplicity, in the sketch, the deformed configuration starts from point O, the starting point of the first element, and arrives at point E', the final point of the Nth element.

During the actuation, each of these N elements undergoes a rotation of an angle φ with respect to the element that precedes. Therefore, the last element of the rod will be rotated by an angle that is $N\varphi$ with respect to the horizontal; the upper head, in the final position (segment A'B'), will form the same angle of value $N\varphi$ with respect to initial position (segment AB).

The CA segment (drawn in blue) represents the antagonist wire in the rest condition; once the deformation has occurred, this wire changes its configuration, moving to the CA segment and exerting the antagonistic force F_{ant}.

The DB segment (drawn in red) represents the SMA wire in the rest condition; once the deformation has taken place, this wire also changes its configuration, leading to it coinciding with the DB segment and exerting the F_{SMA}.

The length a represents the distance between the point of application of the force exerted by the antagonist wire and the axis of the actuator at rest. The length r represents the distance between the point of application of the force exerted by the SMA wire and the axis of the actuator at rest.

It should be noted that the configuration shown in Figure 5 applies both in *Case A* (main directions) and in *Case B* (secondary directions). In the latter, parameters a and r and the value of the multiplicative coefficient to be given to the forces involved are different, obtained simply from the different geometry of the system. In *Case A*, the value of the single antagonistic force must be multiplied by two, and in *Case B*, it is necessary to multiply the value of the single SMA force by two.

The device reaches the equilibrium configuration when the sum of the moment of the antagonistic force and of the moment of the elastic reaction of the central rod balances the moment of the SMA force, with respect to the module base (Figure 6). This will occur for a particular angle φ; in fact, the three moments all depend, obviously in a different way, on this angle. Once the φ equilibrium value is obtained, it is possible to determine the elastic deformation of the system, and therefore, its effective working space.

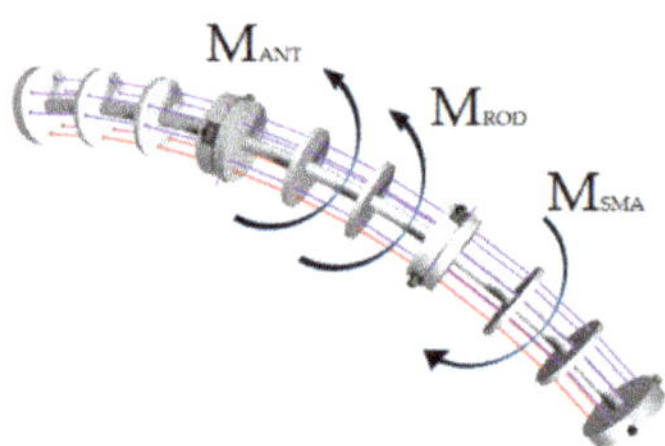

Figure 6. Moments acting on the actuator in the equilibrium condition.

The three considered moments have a different dependence on the deformation angle; therefore, in the following, the procedure that led to the relationships between the forces and the deformation is presented.

Moment of the SMA Force (One Wire)

In our model, the value of the force exerted by the SMA wire is considered to be constant; in fact, it is assumed that, after a short transient, the wire completes the phase transition and is able to exert the nominal force (value given by the manufacturer). The value of the arm of the force applied by the

SMA wire depends by the deformation angle φ. The moment exerted by a SMA wire with respect to the actuator end point is described by the following formula:

$$M_{SMA} = F_{SMA} \cdot r \cdot \cos\left[a\tan\left(\frac{L \cdot \sum_{i=1}^{N} \sin(i\varphi)}{\left(L \cdot \sum_{i=1}^{N} \cos(i\varphi)\right) - r \cdot \sin(N\varphi)}\right)\right] \tag{1}$$

Moment of the Antagonistic Force (One Wire)

The moment of an antagonist wire with respect to the actuator base is described by the following formula:

$$M_{ANT} = F_{ANT} \cdot L_{ANT} = \varepsilon \cdot EA \cdot L_{ANT} \tag{2}$$

where A is wire section area, E is wire's Young modulus and

$$\varepsilon = \frac{L_{final} - L_{start}}{L_{start}} \tag{3}$$

is the longitudinal deformation of the wire.

While the section of the wire and the linear elastic modulus of the material can be considered constant parameters, the value of the longitudinal deformation varies with the angle of deformation because the final length of the wire is a parameter that depends on the angle φ.

As can be seen, in this case, the deformation angle does not only influence the arm of the force with respect to the final point of the rod but also affects the value of the antagonist force applied by the wire, so the final equations are the following:

$$F_{ANT} = \frac{EA}{L} \cdot \left[\sqrt{\left(L \cdot \sum_{i=1}^{N} \cos(i\varphi) + a \cdot \sin(N\varphi)\right)^2 + \left(a - a \cdot \cos(N\varphi) + L \cdot \sum_{i=1}^{N} \sin(i\varphi)\right)^2} - L\right] \tag{4}$$

$$L_{ANT} = a \cdot \cos\left[a\tan\left(\frac{L \cdot \sum_{i=1}^{N} \sin(i\varphi)}{L \cdot \sum_{i=1}^{N} \cos(i\varphi) - r \cdot \sin(N\varphi)}\right)\right] \tag{5}$$

Moment of the Elastic Reaction of the Central Rod

The stiffness of the spring that represents the flexibility of the central rod depends on the material (Young modulus), on the section area, and on the length of the rod itself according to the following relation, so the final moment of the central rod can be described using the following equation:

$$M_{rod} = \frac{E_r I}{L} N\varphi \tag{6}$$

where

E_r is the Young modulus of the material constituting the rod;

I is the moment of inertia of the rod section;

L is the length of the rod.

Model Results

By way of example, the results of the model are presented in the present paragraph in the case of a finger consisting of three modules 40 mm long each, placed in series and having a Polytetrafluoroethylene (PTFE) central rod, with the characteristics summarized in Table 1.

Table 1. Actuator main data.

Central rod diameter (mm)	2.6
Total actuator length (mm)	120
Young's modulus PTFE (MPa)	450
SMA wires' positioning radius (mm)	5
SMA wire diameter (µm)	250
SMA wire shortening (%)	5
SMA Young modulus in martensitic phase (GPa)	28
SMA Young modulus in austenitic phase (GPa)	75

When the actuator reached the theoretical equilibrium condition, the model gave the results summarized in Table 2.

Table 2. Model results.

	Main Direction	Secondary Direction
Total SMA moment (Nm)	0.0424	0.0454
Total antagonistic moment (Nm)	0.0400	0.0440
Total central rod moment (Nm)	0.0024	0.0014
Total deformation angle Nf (rad)	0.028	0.017

Figure 7 shows the final deformed configuration of the finger, both along the main directions (dotted line) and the secondary ones (solid line), considering the actuator axis in a rest condition laying along the Z-axis. This graph represents the deformation of the actuator axis on the deformation plane; this plane contained the axis of the actuator, and the forces exerted by the SMA wires and by the antagonist wires.

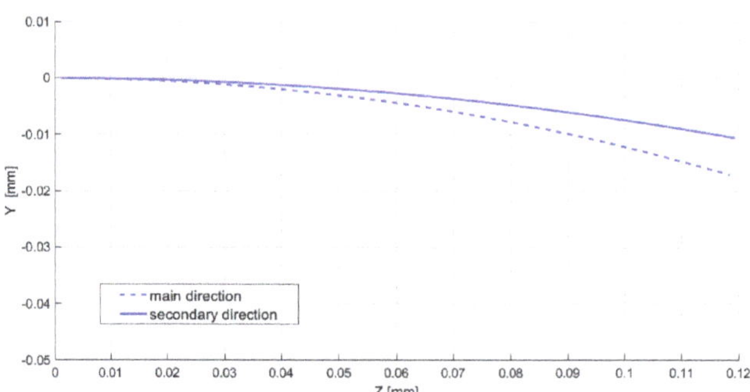

Figure 7. Axis of central rod of the actuator in equilibrium condition.

Figure 8 shows the projection of the deformation on a plane perpendicular to the actuator axis (X-Y).

As can be seen, and with reference to Figure 2, the greater deformations occurred along the three main directions, which, starting from the X-axis, were arranged to form three 120° angles between them. Instead, along the three secondary directions that were also arranged at 120° between them but 60° out of phase with the previous ones, the deformations were smaller. This was due to the fact that along the secondary directions, although there were two SMA wires activated at a time and there was only one antagonist wire, the arm of action of the force exerted by the two SMA wires was very small with respect to the arm of the force exerted by the antagonist wire. On the contrary, in the main directions, only one SMA wire was operated, but the arm of the force exerted by the SMA wire was bigger.

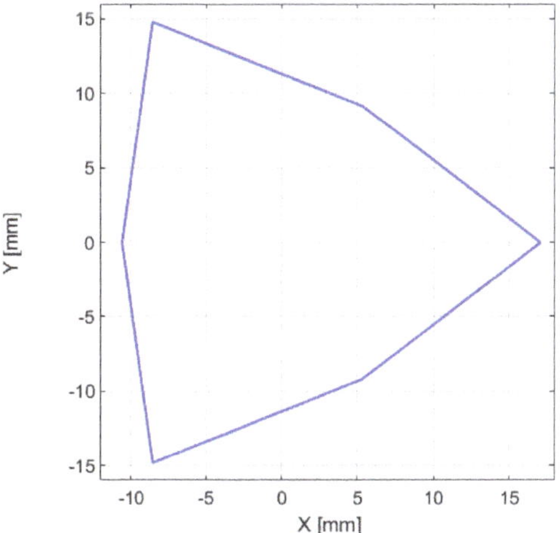

Figure 8. XY projection of the workspace.

4. Prototypes Designs

In Table 3, the characteristics of the materials of three different realized prototypes are presented, and Figure 9 shows the prototypes.

Table 3. Prototypes characteristics.

	Material	Young's Modulus (MPa)	Melting Temperature (°C)
(A)	Nylon	800	120
(B)	PTFE	800	250
(C)	PEEK1000/LIM	4400/50	340

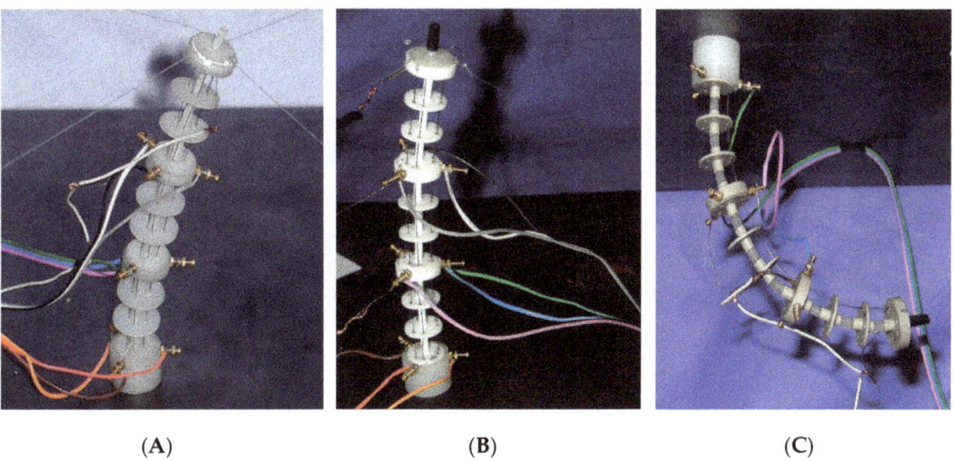

(A) **(B)** **(C)**

Figure 9. Prototypes A with nylon central rod (**A**), B with PTFE central rod (**B**), and C with LIM central rod and PEEK1000 disks (**C**).

Due to a Young's modulus of the nylon central rod of about 800 MPa, prototype A was characterized by a quite good flexibility. The advantage of this solution is that the material is inexpensive and its machine tooling process is simple, but the disadvantage is that its melting temperature, around 120 °C, is close to the temperatures reached by the activated hot wires. Some results of experimental testing on this prototype can be found in Maffiodo and Raparelli [32]. Prototype B, which had a central module made of PTFE, was then built. This material is more expensive, in particular with regard to tool machining. The advantages, however, are a similar Young's modulus, excellent flexibility of the structure, and a decidedly higher melting temperature.

Prototype C was created with the idea of separating the need to have a high melting temperature of the areas in contact with the SMA hot wires and the need to have a high flexibility of the central shaft. Then, the bases and intermediate disks were made with a technopolymer, PEEK1000, with a high Young's modulus and with a high melting temperature, while the flexible elements were made of silicone rubber (LIM) with a low Young's modulus.

5. Experimental Tests Bench, Procedure, and Results

The experimental set up, sketched in Figure 10, is composed of a flat square plate with four vertical rods at its edges. The prototype is clamped in the middle of the plate and four low-friction nylon wires are fixed to the actuator upper end. The opposite end of the wire is connected to a load with the aim of tightening the wire itself and to calculate the actuator end position in space using simple algebraic and trigonometric calculations.

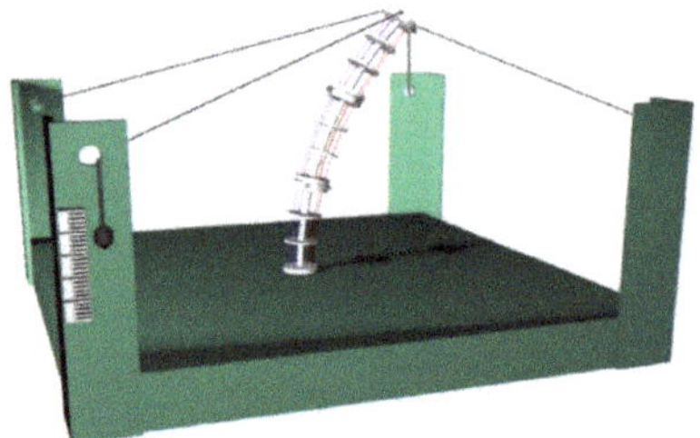

Figure 10. Sketch of the test bench.

The test aimed to evaluate the actuator displacement along the three main directions and the three secondary directions (cf. Figure 2) during heating and cooling sequences. The test was carried out with a step supply current of 1 A for the heating phase (duration 30 s). The activation sequence for each direction was the following: activation of the lower module, activation of the central module, activation of the higher module. Cooling was in still air. Each sequence was repeated at least three times. The power supply was 2.8 W.

The tests performed on prototype A highlighted its stable and repeatable behavior. The 3D workspace, visible in Figure 11, was sufficiently large. The melting temperature close to the temperatures reached by the SMA wire in the hot phase was an important limit of the prototype, which after a relatively low number of cycles had been damaged in these areas. In a first instance, a local insulation can be added, but design changes have to be implemented.

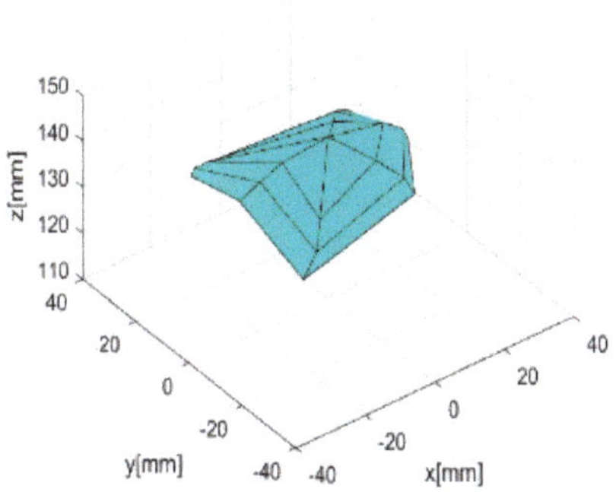

Figure 11. 3D workspace of prototype A.

Prototype B, despite having mechanical characteristics similar to prototype A, had a slightly lower working space. It was observed that this difference can be attributed to a different pretensioning of the antagonist wires in the two prototypes. This observation has led to the need to investigate in more depth the contribution of inactive antagonistic SMA wires, which seemed to significantly limit the deformation of the entire actuator. Figure 12 shows the comparison of the tests obtained on prototype B in the case of the presence of antagonist wires (a), and in the absence of antagonist wires (b). Obviously, this situation, which was artificially generated by disassembling the antagonist wires, is not compatible with the normal use of the device, but is used only for the purpose of investigation. In the result obtained, namely a considerable increase in the working space in the second case, encourages the realization of a new device in which an additional locking/unlocking device allows for the disabling of the antagonist wire when necessary.

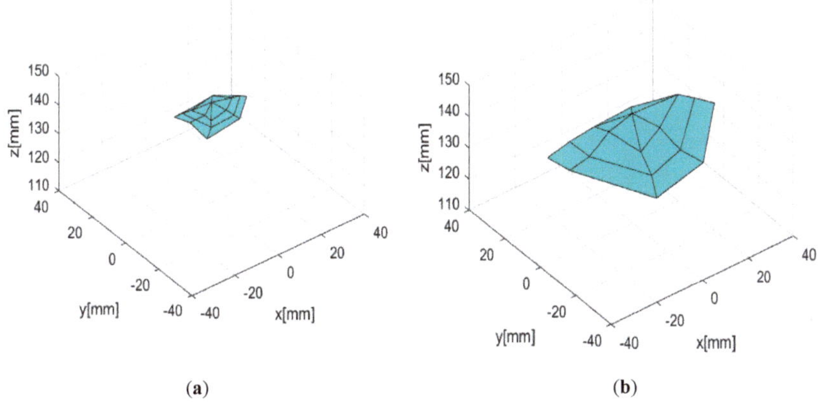

(a)　　　　　(b)

Figure 12. (a) 3D workspace of prototype B, and (b) 3D workspace of prototype B without antagonistic wires.

Figure 13 shows the comparison of the two different workspaces on an X-Y projection.

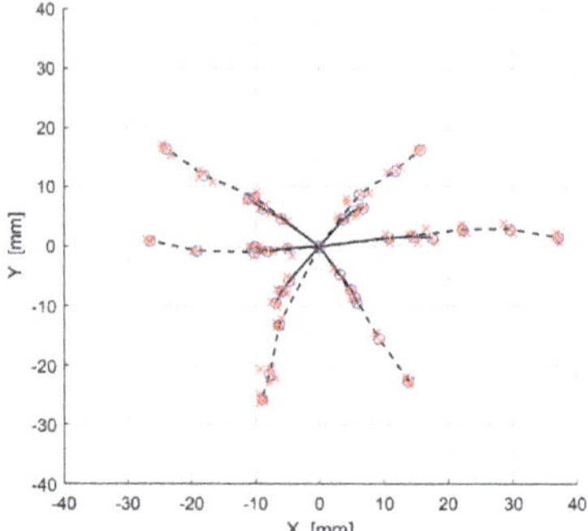

Figure 13. X-Y projection of the workspaces for prototype B (cross: experimental, circle: mean values; solid line: with antagonistic wires, dotted line: without antagonistic wires).

Figure 14 shows the results obtained with prototype C, which was also tested in the case of the presence of antagonistic wires (a) and in the absence of antagonist wires (b). It can be observed that, also in this case, the 3D workspace, already quite large in case (a), significantly increased in case (b). The device, having good thermomechanical characteristics, was resistant to prolonged use over time. However, silicone rubber was shown to be insufficient to exercise the correct elastic recall for the recovery of undeformed conditions at rest.

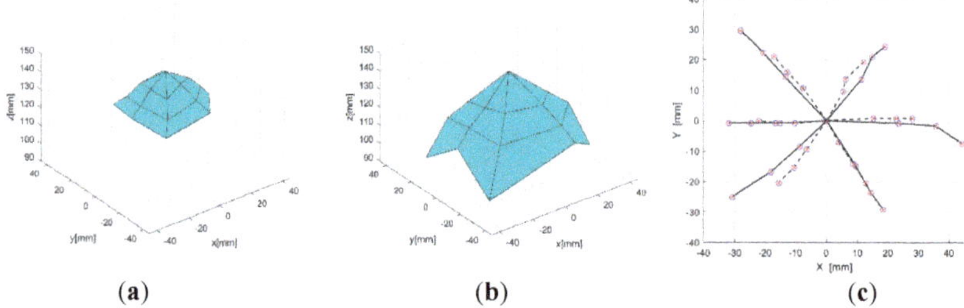

(a)	(b)	(c)

Figure 14. 3D work space of prototype C: (**a**) complete; (**b**) without antagonistic wires; (**c**) X-Y projection of the workspaces, (cross: experimental, circle: mean values; solid line: with antagonistic wires, dotted line: without antagonistic wires).

6. New Prototype D

Based on the results of the experimental tests and the observation that the absence of antagonistic wires would benefit the width of the workspace, a fourth prototype was designed. In this new device, the wires could be blocked for the active phase and released during inactive phases, so as not to work against the active wire. It was, however, necessary that a bias force was present for the return to the undeformed condition of the device. This new device must be as simple and compact as possible, even if the introduction of these new features necessarily involves a greater complexity and an increase in the number of components.

The structure of each module of the new prototype was similar to that of the previous prototypes, except for the addition of the locking/unlocking device housed in a base seat. Figure 15a shows the assembly of the prototype. As can be seen, the addition of a locking/unlocking device has been added below each "traditional" module. In order to facilitate the connection of the wire ends to the cylinder, the distance between the holes for the passage of one wire was modified with respect to the previous solutions (25° instead of 40°). The materials chosen for the new prototype were the following: PTFE for the module, PEEK100 for the base seat, and PES (Polysulfone) for the cylinders.

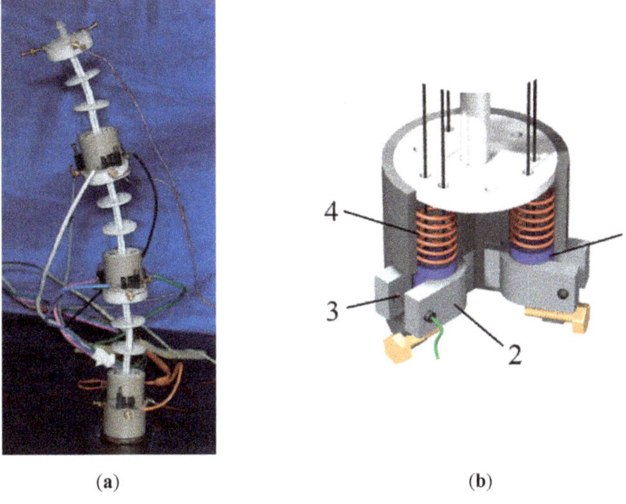

(a)   (b)

Figure 15. (a) New prototype D. (b) Detail of activation/deactivation device on the module base.

Figure 15b shows the detail of the locking/unlocking device. Both ends of each of the three SMA wires were connected to a cylinder (1) located below the lower base. A cylinder (1) was placed inside a gripper (2), which was driven by an additional SMA wire (3). When a current was sent to the additional SMA wire of the gripper, the cylinder was locked and could not move vertically anymore. Subsequently, current was sent to the corresponding SMA wire of the module, which would consequently flex along a main direction. During this phase, the cylinders of the two inactive SMA wires were free to slide vertically because the corresponding grippers were also inactive. Therefore, outside of the friction between the cylinder and gripper, the active wire was not impeded by antagonist wires. When the power supply was interrupted, the clamping action also stopped and the cylinder was free to move again. A similar behavior occurred for the activation along the secondary directions. The bias force for the cylinders and rod was guaranteed by using appropriately sized springs (k = 2.28 N/mm) (4).

A set of preliminary tests of the device has been carried out and the results are shown in Figure 16 (continuous black line). The projection in the XY plane of the work space is here compared with the results obtained with prototype B, which had a central rod made of the same material. Compared to the solution with antagonistic wires (green solid line), the results of the new prototype were better. Obviously, the presence of sliding friction in the contact between the gripper and cylinder meant that it was not possible to reach the deformation obtained by completely eliminating the antagonist wires from prototype B (green dotted line).

The difference in orientation of the flexion directions of the new prototype with respect to prototype B could be attributed to the sum of two problems related to the experimental procedure: the pretensioning of the SMA wires and the difficulty in orienting the two devices exactly in the same way with respect to the test bench.

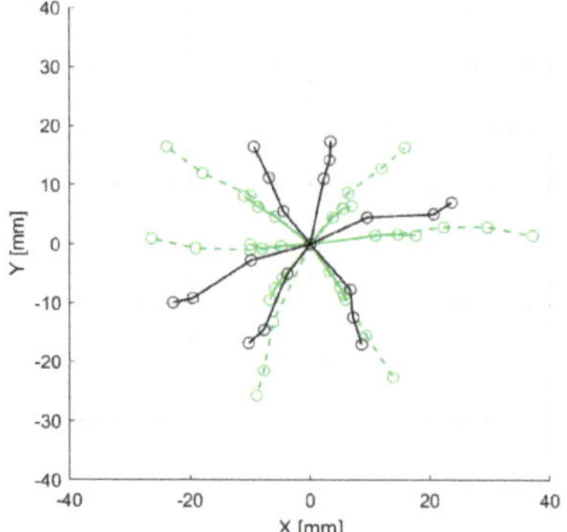

Figure 16. X-Y projection of the workspace with prototype D in black and prototype B in green (solid line: with antagonistic wires, dotted line: without antagonistic wires).

7. Discussion

The need to create flexible and versatile gripping hands has led to the creation of a flexible module operated by SMA wires, which can be the basic "brick" with which to assemble many different solutions.

A simple mathematical model was implemented with the aim of theoretically describing the behavior of the device and evaluating its working space.

Based on the results obtained, three devices with the same geometry were developed using three different materials. Experimental tests were conducted on these three prototypes, which allowed the evaluation of the workspace and highlighted some advantages and disadvantages. The working space of prototype A proved to be adequate, but the module was damaged at some points due to the low melting temperature of the nylon. With prototype B, this limit was avoided because the PTFE the central module was made of has a high melting temperature. However, despite a similar nominal elastic module, it showed a reduced working space. Therefore, by accepting an increased difficulty in assembling the module, a solution was conceived, thanks to which large working spaces were obtained and no overheating problems were encountered. This solution (prototype C) involved the use of different materials for the disks and for the central shaft, respectively in PEEK1000 and LIM, but a limitation was introduced by the difficulty in returning to the undeformed condition and a consequent non-repeatability of the tests.

Prototypes B and C were also tested for the evaluation of the influence of the presence of inactive SMA on the work space amplitude, conducted by simply removing these wires and verifying the bending of the central module along the main and secondary directions in the presence of one or two active wires. The results of these tests led to the design of a further prototype. Here, in prototype D, each module had a device for locking/unlocking the SMA wires, governed in turn by an additional SMA wire. Results from the tests on this last prototype are encouraging because its workspace was greater than the workspace of the prototype without a locking/unlocking device.

However, further improvements could be obtained by reducing the sliding friction observed in the new device and by solving some pre-tension and alignment issues. The former could be reduced by taking into consideration other materials for the realization of the gripper/cylinder coupling having a lower friction coefficient, and having greater care for surface roughness. The latter are being resolved through the implementation of appropriate assembly equipment.

Author Contributions: Conceptualization, D.M. and T.R.; methodology, D.M. and T.R.; software, D.M.; validation, D.M.; writing—original draft preparation, D.M.; writing—review and editing, D.M.; supervision, T.R.

Funding: This research received no external funding.

Conflicts of Interest: The authors declare no conflict of interest.

References

1. Hodgson, D.E.; Wu, M.H.; Biermann, R.J. Properties and Selection: Nonferrous Alloys and Special-Purpose Materials. In *ASM Handbook*; ASM International: Metals Park, OH, USA, 1990; Volume 2, pp. 897–902.
2. Raparelli, T.; Beomonte Zobel, P.; Durante, F. Mechanical design of a 3-dof parallel robot actuated by smart wires. In Proceedings of the 2nd European Conference on Mechanism Science, Cassino, Italy, 17–20 September 2008; pp. 271–278. [CrossRef]
3. Raparelli, T.; Zobel, P.B.; Durante, F. A proposed methodology for the development of microgrippers: An application to a silicon device actuated by shape memory alloy wires. *Int. J. Mech. Eng. Technol.* **2018**, *9*, 235–249.
4. Lee, K-T.; Lee, G-Y.; Choi, J-O.; Wu, R.; Ahn, S-H. Design and Fabrication of a Smart Flexible Structure using Shape Memory Alloy Wire (SMA). In Proceedings of the 2010 3rd IEEE RAS & EMBS International Conference on Biomedical Robotics and Biomechatronics, Tokyo, Japan, 26–29 September 2010; pp. 599–603.
5. Wang, Z.; Hang, G.; Li, J.; Wang, Y.; Xiao, K. A micro-robot fish with embedded SMA wire actuated flexible biomimetic fin. *Sens. Actuators A* **2008**, *144*, 354–360. [CrossRef]
6. Fukuda, T.; Hosokai, H.; Kikuchi, I. Distributed type of actuators by shape memory alloy and its application to underwater mobile robotic mechanism. In Proceedings of the IEEE International conference on Robotics and Automation, Cincinnati, OH, USA, 13–18 May 1990; pp. 1316–1321. [CrossRef]
7. Maffiodo, D.; Raparelli, T. Three-fingered gripper with flexure hinges actuated by shape memory alloy wires. *Int. J. Autom. Technol.* **2017**, *11*, 355–360. [CrossRef]
8. Giataganas, P.; Evangeliou, N.; Koveos, Y.; Kelasidi, E.; Tzes, A. Design and experimental evaluation of an innovative SMA-based tendon-driven redundant endoscopic robotic surgical tool. In Proceedings of the 19th Mediterranean Conference on Control & Automation, Corfu, Greece, 20–3 June 2011; pp. 1071–1075.
9. Duerig, T.; Pelton, A.; Stöckel, D. An overview of nitinol medical applications. *Mater. Sci. Eng. A* **1999**, *273–275*, 149–160. [CrossRef]
10. Petrini, L.; Migliavacca, F. Biomedical Applications of Shape Memory Alloys. *J. Metall.* **2011**, *2011*, 501483. [CrossRef]
11. Song, G.; Kelly, B.; Agrawal, B.N. Active position control of a shape memory alloy wire actuated composite beam. *Smart Mater. Struct.* **2000**, *9*, 711–716. [CrossRef]
12. Lima, W.M.; Araujo, C.J.D.; Valenzuela, W.A.V.; Rocha Neto, J.S.D. Control of strain in a flexible beam using Ni-Ti-Cu shape memory alloy wire actuators. *J. Braz. Soc. Mech. Sci. Eng.* **2012**, *34*, 413–422. [CrossRef]
13. Ikuta, K. Micro/miniature shape memory alloy actuator. In Proceedings of the IEEE International Conference on Robotics and Automation, Cincinnati, OH, USA, 13–18 May 1990; pp. 2156–2161. [CrossRef]
14. Jani, J.M.; Leary, M.; Subic, A.; Gibson, M.A. A review of shape memory alloy research, applications and opportunities. *Mater. Des.* **2014**, *56*, 1078–1113. [CrossRef]
15. Shameli, E.; Alasty, A.; Salaarieh, H. Stability analysis and nonlinear control of a miniature shape memory alloy actuatorfor precise applications. *Mechatronics* **2005**, *15*, 471–486. [CrossRef]
16. Choi, S.B.; Han, Y.M.; Kim, J.H.; Cheong, C.C. Force tracking control of a flexible gripper featuring shape memory alloy actuators. *Mechatronics* **2001**, *11*, 677–690. [CrossRef]
17. Maffiodo, D.; Raparelli, T. Resistance feedback of a shape memory alloy wire. *Adv. Intell. Syst. Comput.* **2016**, *371*, 97–104. [CrossRef]
18. Tsugami, Y.; Barbiè, T.; Tadakuma, K.; Nishida, T. Development of Universal Parallel Gripper using Reformed Magnetorheological Fluid. In Proceedings of the 11th Asian Control Conference (ASCC) Gold Coast Convention Centre, Gold Coast, Australia, 17–20 December 2017.
19. Felser, A.; Zieve, P.B.; Ernsdorff, B. Use of Synchronized Parallel Grippers in Fastener Injection Systems. *SAE Technical Paper* **2015**. [CrossRef]
20. Sudsang, A.; Ponce, J. New techniques for computing four-finger force closure grasps of polyhedral objects. In Proceedings of the IEEE International Conference on Robotics and Automation, Nagoya, Japan, 21–27 May 1995.
21. Li, Z.; Hsu, P.; Sastry, S. Grasping and coordinated manipulation by a multifinger robot hand. *Intl. J. Robot. Res.* **1989**, *8*, 33–50.
22. Datta, R.; Pradhan, S.; Bhattacharya, B. Analysis and Design Optimization of a Robotic Gripper Using Multiobjective Genetic Algorithm. *IEEE Trans. Syst. Man Cybern. Syst.* **2016**, *46*, 16–26. [CrossRef]

23. Baliga, U.B.; Winston, S.J.; Sandeep, S. Design Optimization of Power Manipulator Gripper for Maximum Grip Force. *Int. J. Eng. Res. Technol.* **2014**, *3*, 134–140.

24. Jia, Y.; Zhang, X.; Xu, Q. Design and optimization of a dual-axis PZT actuation gripper. In Proceedings of the IEEE International Conference on Robotics and Biomimetics (ROBIO 2014), Bali, Indonesia, 5–10 December 2014; pp. 321–325. [CrossRef]

25. Modabberifar, M.; Spenko, M. A shape memory alloy-actuated gecko-inspired robotic gripper. *Sens. Actuators A Phys.* **2018**, *276*, 76–82. [CrossRef]

26. Schulte, H.F., Jr. The characteristics of the McKibben artificial muscle. In *The Application of External Power in Prosthetics and Orthotics*; National Academy of Sciences-National Research Council: Washington, DC, USA, 1961; pp. 94–115.

27. Inoue, K. Rubbertuators and applications for robots. In *Robotics Research: The 4th International Symposium*; Bolles, R., Roth, B., Eds.; MIT Press: Cambridge, MA, USA, 1988; pp. 57–63.

28. Ferraresi, C.; Franco, W.; Quaglia, G. A novel bi-directional deformable fluid actuator. *Proc. Inst. Mech. Eng. Part C* **2014**, *228*, 2799–2809. [CrossRef]

29. Yang, K.; Gu, C.L. A compact and flexible actuator based on shape memory alloy springs. *J. Mech. Eng. Sci.* **2008**, *222*, 1329–1337. [CrossRef]

30. Torres-Jara, E.; Gilpin, K.; Karges, J.; Wood, R.J.; Russ, D. Compliant Modular Shape Memory Alloy Actuators. *IEEE Rob. Autom. Mag.* **2010**, *17*, 78–87. [CrossRef]

31. Maffiodo, D.; Raparelli, T. Design and realization of a flexible finger actuated by shape memory alloy (SMA) wires. *Int. J. Appl. Eng. Res.* **2017**, *12*, 15635–15643.

32. Maffiodo, D.; Raparelli, T. Experimental testing of a modular flexible actuator based on SMA wires. *Int. J. Appl. Eng. Res.* **2018**, *13*, 1465–1471.

 *machines*

Article

Design and Prototyping of Miniaturized Straight Bevel Gears for Biomedical Applications

Antonio Acinapura [1],*, Gionata Fragomeni [2], Pasquale Francesco Greco [3], Domenico Mundo [1], Giuseppe Carbone [1] and Guido Danieli [3]

[1] DIMEG, University of Calabria, 87036 Rende (CS), Italy; domenico.mundo@unical.it (D.M.); giuseppe.carbone@unical.it (G.C.)
[2] Magna Graecia University of Catanzaro, 88100 Germaneto (CZ), Italy; gionata.fragomeni@unical.it
[3] Calabrian High Tech—CHT S.r.l., 87036 Rende (CS), Italy; pasquale.greco@chtsrl.com (P.F.G.); guido.danieli@unical.it (G.D.)
* Correspondence: antonio.acinapura@unical.it; Tel.: +39-0984-494159

Received: 8 May 2019; Accepted: 31 May 2019; Published: 4 June 2019

Abstract: This paper presents a semi-automated design algorithm for computing straight bevel gear involute profiles. The proposed formulation is based on the Tredgold approximation method. It allows the design of a pair of bevel gears with any desired number of teeth and relative axes inclination angles by implementing additive manufacturing technology. A specific case study is discussed to calculate the profiles of two straight bevel gears of a biomedical application. Namely, this paper illustrates the design of the bevel gears for a new laparoscopic robotic system, EasyLap, under development with a grant from POR Calabria 2014–2020 Fesr-Fse. A meshing analysis is carried out to identify potential design errors. Moreover, finite element-based tooth contact analysis is fulfilled for determining the vibrational performances of the conjugate tooth profiles throughout a whole meshing cycle. Simulation results and a built prototype are reported to show the engineering feasibility and effectiveness of the proposed design approach.

Keywords: bevel gears; gear design; Tredgold; numerical simulations

---

1. Introduction

Over the last decades, robotic systems have improved the performance in terms of time-consuming and quality of many biomedical applications. In particular, endoscopic surgical operations requiring accurate movements and minimal invasion may get significant advantages from robotic assistance. Laparoscopy was firstly promoted as a diagnostic modality [1,2] and later it was used for operations in the fields of gynecology, urology, and cardiac surgery [3–5]. AESOP, the automated endoscope system for optimal positioning, represented the first solution eliminating the need for one additional person at the operating table. It is a computer-controlled robot positioner that holds the laparoscope and is moved by means of surgeon's foot and hand controllers [6]. Another kind of computer-assisted system has been developed to give assistance in endoscopic coronary artery surgical procedures. The so-called ZEUS system is a 5-degrees of freedom (DOF) voice-controlled robotic surgical system [7], which allows duplicating the DOFs of a standard endoscopic instrument controlled by the hand of a surgeon. Additionally, the da Vinci robotic surgical system adopts the principle of a master–slave manipulator, in which the handle of the instrument and its top part are physically divided [8]. Although da Vinci and ZEUS provided certain advantages like high visual magnification, movement scaling, tremor filtering, and dexterity, some drawbacks must be underlined, such as interference among the robotic arms, excessive encumbrances, and lack of tactile feedback. These drawbacks motivated the exploration of other design solutions such as the EasyLap robotic system [9], which uses instrumentation already available in hospitals.

The aim of this work is to improve the performances of the EasyLap robotic system for laparoscopic surgery operations [9]. It is made of four to five arms, which are fixed on a common base frame, each being passively self-balanced with five DOFs, while the sixth is supplied by the instrument in use turning about its axis. Four arms are usually actuated, while one is dedicated to the optics and is automatically moved, pointing to the region in which the instruments are manipulated by the doctor. The attention is focused on the design of the adaptor that handles the movements of the traditional surgical instrument that must turn along its axis to modify the plane of opening of the forceps, which needs a particular attention being the most significant part, especially for endoscopic surgery operations.

One of the key features of the EasyLap robotic system is the possibility to fully detach the sterile component (holding the forceps) from the non-sterile part (containing the motors, force sensor, and electronics). For this purpose, it is necessary to keep the motors parallel and, at same time, it is necessary to provide a low-cost transmission solution for the rotation of the end-effector with specific small relative axes inclination angles and reducing ratios of approximately 0.5. A careful search found that there is no off-the-shelf solution being able to satisfy both required transmission and geometric constraints. Accordingly, there has arisen a need to design a customized bevel gear solution, which can satisfy all the design constraints such as inclination/transmission angles with a low-cost and compact solution. Additionally, standard gear design solutions fail to provide a customized design with the desired design characteristics and performance, especially for reducing the encumbrance and, at the same time, ensure good vibrational behavior of the end-effector.

The traditional manufacturing processes are mainly cutting methods that strictly influence the tooth profile of a gear and, thus, its performance. Face milling and face hobbing are two of the main generating methods. However, they do not guarantee customizable control of the tooth profile and good vibrational performances when applying small loads [10]. Accordingly, a specific design methodology is herewith proposed. This design methodology is based on an extension of the Tredgold approximation method [11]. It is intended for miniaturized straight bevel gears with any desired number of teeth and relative axes inclination angles.

The proposed design method assumes the use of an additive manufacturing technology that was not considered in conventional design methods such as those proposed in literature [12,13]. The proposed method implements a semi-automated pre-processing tool to generate a point-based description of the tooth flank surfaces and then to create a desired finite element (FE) model of the gear pair. The FE model has been used to compute a tooth contact analysis (TCA) of the mating gears in order to determine their vibrational performances, which are extremely significant for the considered biomedical application. The TCA is performed by means of nonlinear FE simulations in the NX Nastran (SOL 601) environment. Simulation results are discussed to attest to the engineering feasibility and effectiveness of the proposed design approach. A 3D-printed prototype of the designed mating elements has been manufactured and mounted on the end-effector of the EasyLap.

2. Tredgold-Based Tooth Profile Generation

Bevel gears are used when the motion has to be transmitted between intersecting shaft axes. In this work, straight bevel gears with involute tooth profiles were considered in order to actuate the end-effector of the robotic system EasyLap when used for laparoscopic surgery operations. Since encumbrances must be minimized, it is necessary to perform a careful design of the mating gears. Small gears and small shaft angle represent the main topics of the proposed gear design.

Traditionally, bevel gears are manufactured by using cutting methods such as gear hobbing, bevel gear generators, or CNC milling. These methods start from a bulk workpiece and remove material to generate a correctly profiled tooth form [14–16]. These manufacturing processes do not allow for an easy customization and miniaturization. Other manufacturing methods can be based on die casting or injection molding [17]. These methods are suitable only for large production lots, since they require an expensive mold design and production. New production processes such as 3D printing have been changing the time-to-market of industrial products but also the design process. In this work, the use of 3D printing techniques is taken into account for modifying the generating method that is used to design a desired straight bevel gear pair. This method can be very fast and cost effective for small production lots of small size plastic gears, such as those considered for the EasyLap robotic system. However, it is necessary to fully design the bevel gear mating profiles in terms of a cloud of points to be provided as input of a 3D printer. Accordingly, this paper proposes a specific semi-automated design algorithm for computing straight bevel gear involute profiles.

The Tredgold approximation method [11] is usually adopted to study bevel gears, reducing the problem to that of ordinary spur gears. In the proposed approach, the aim is not only to study bevel gear problems but also to design the bevel gear pair starting from its equivalent spur one. As seen in Figure 1, r_1, and r_2 are the pitch radii, respectively, for the pinion and the gear. The angles γ_1 and γ_2 are defined as pitch angles and their sum is equal to the shaft angle Σ.

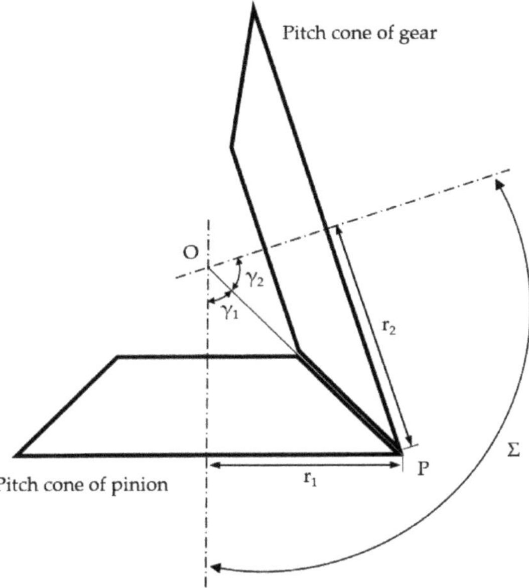

Figure 1. Pitch cones of the gear pair elements.

The gear ratio is defined in the same manner as for spur gears:

$$\tau = \frac{\omega_1}{\omega_2} = \frac{r_2}{r_1} = \frac{z_2}{z_1}, \tag{1}$$

where ω and z are, respectively, angular speed and number of teeth of the considered gear. Since the distance OP can be written as

$$OP = \frac{r_1}{\sin \gamma_1} = \frac{r_2}{\sin \gamma_2}, \tag{2}$$

then combining them gives

$$\sin \gamma_1 = \frac{r_1}{r_2} \sin \gamma_2 = \frac{r_1}{r_2} \sin(\Sigma - \gamma_1) = \frac{r_1}{r_2}(\sin \Sigma \cos \gamma_1 - \sin \gamma_1 \cos \Sigma). \tag{3}$$

Rearranging Equation (3) gives

$$
\begin{cases}
\tan \gamma_1 = \dfrac{\sin \Sigma}{\frac{r_2}{r_1}+\cos \Sigma} = \dfrac{\sin \Sigma}{\frac{z_1}{z_2}+\cos \Sigma} \\[2ex]
\tan \gamma_2 = \dfrac{\sin \Sigma}{\frac{r_1}{r_2}+\cos \Sigma} = \dfrac{\sin \Sigma}{\frac{z_2}{z_1}+\cos \Sigma}
\end{cases} .
\tag{4}
$$

The pitch radius r_i for the i-th gear is defined as

$$
r_i(w) = m(w)\frac{z_i}{2},
\tag{5}
$$

where $m(w)$ is the module. Considering the bevel gears, m varies as a function of the face width of the considered gear. Focusing the attention on Equations (4) and (5), once the shaft angle, the number of teeth, and the function of the module are fixed, it is possible to obtain the pitch cones of the gear pair elements.

The Tredgold approximation method allows starting from the pitch back-cone representation of a bevel gear pair in order to build an equivalent problem of spur gears. As long as the gear is made up of eight or more teeth, the method is accurate enough for practical purposes [12]. According to the Tredgold method, an equivalent spur gear is built whose pitch radius r_e is equal to the back-cone radius, such as the radius of the cone whose elements are perpendicular to those of the pitch cone at the largest end of the teeth. Once the equivalent spur gears are obtained, the tooth profiles can then be defined. The results obtained by tooth contact analysis of the equivalent gears correspond closely to those of the bevel gears.

The geometrical relations that occur between the bevel and spur gears are

$$
r_{eq,i}(w) = \frac{r_i(w)}{\cos \gamma_i},
\tag{6}
$$

$$
z_{eq,i}(w) = \frac{2\,\pi\, r_{eq,i}(w)}{p(w)} = \frac{z_i(w)}{\cos \gamma_i},
\tag{7}
$$

where $r_{eq,i}$ and $z_{eq,i}$ are, respectively, the pitch radius and the number of teeth, usually a non-integer, of the i-th equivalent spur gear; p is the circular pitch of the bevel gear measured at the width w of the gear. In order to make the following discussion easier to understand, the procedure will be shown for a single gear pair element and for a fixed value of the module m, which would be repeated for a discrete number n of intermediate values between m_{bc} (module at the back cone) and m_{fc} (module at the front cone).

Once the geometrical parameters of the equivalent gears are obtained, it is possible to focus the attention on the 2D design of the involute profiles using the following Cartesian equations:

$$
\begin{cases}
x_{eq} = r_{b,eq}(\sin \theta - \theta \cos \theta) \\
y_{eq} = r_{b,eq}(\cos \theta + \theta \sin \theta)
\end{cases} ,
\tag{8}
$$

where $r_{b,eq}$ is the radius of the fundamental circle of the equivalent gear and θ is the involute angle that varies between zero and θ_t.

$$
\theta_t = \sqrt{\left(\frac{r_{t,eq}}{r_{b,eq}}\right)^2 - 1},
\tag{9}
$$

where θ_t is the angle at which the involute reaches the addendum circle and $r_{t,eq}$ is its radius. It is usually useful to adopt the polar coordinates (ρ,ϕ) where

$$
\begin{cases}
\rho = \sqrt{\left(x_{eq}^2 + y_{eq}^2\right)} \\
\phi = \operatorname{atan}\!\left(\frac{x_{eq}}{y_{eq}}\right)
\end{cases} .
\tag{10}
$$

A significant value that will be useful for 3D parameterization is ϕ_p, which is the value of ϕ at θ_p, where the involute meets the pitch line. The 3D parameterization of the single tooth of the bevel gears is shown in Equation (11):

$$\begin{cases} x = R\sin\!\left(\dfrac{\delta}{\cos\gamma}\right)\cos\gamma \\ y = R\cos\!\left(\dfrac{\delta}{\cos\gamma}\right)\cos\gamma \\ z = r_{vag} + r_{eq}\tan\gamma - R\sin\gamma \end{cases} \quad , \tag{11}$$

where R and δ are the polar coordinates, while r_{vag} is the radius of the virtual auxiliary gear. The latter should mesh with the equivalent spur gear, but it is only used to define the z-coordinate of the gear flank parameterization. They can be defined as reported in Equations (12)–(14).

$$R = \begin{cases} r_{f,eq} = r_{eq} - 1.25\,m & \text{root part} \\ \rho & \text{active flank part} \\ r_{t,eq} = r_{eq} + m & \text{top part} \end{cases} \quad , \tag{12}$$

$$\delta : \begin{cases} -\dfrac{\pi}{z_{eq}} \le \delta \le -\dfrac{\pi}{2\,z_{eq}} - \left(\phi_p - \phi\right) & \text{left root part} \\ -\dfrac{\pi}{2\,z_{eq}} - \left(\phi_p - \phi\right) < \delta \le -\left(\phi_p - \phi\right) & \text{left active flank part} \\ -\left(\phi_p - \phi\right) \le \delta \le \left(\phi_p - \phi\right) & \text{top flank part} \\ \left(\phi_p - \phi\right) \le \delta < \left(\phi_p - \phi\right) + \dfrac{\pi}{2\,z_{eq}} & \text{right active flank part} \\ \dfrac{\pi}{2\,z_{eq}} + \left(\phi_p - \phi\right) \le \delta \le \dfrac{\pi}{z_{eq}} & \text{right root part} \end{cases} \quad , \tag{13}$$

$$r_{vag} = r_{eq}\sin\gamma. \tag{14}$$

3. The Proposed Semi-Automated Design Method

The proposed semi-automated process to generate the design of straight bevel gears and the contact performance of the prescribed gear pair can be divided into three main phases, as described in Figure 2.

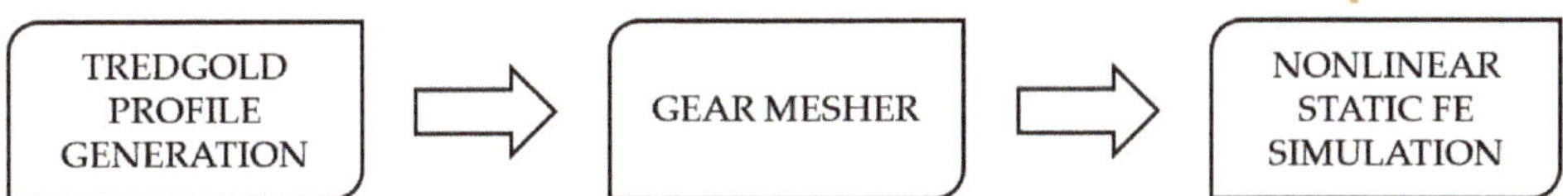

Figure 2. Logical sequence of the proposed semi-automated design method.

The Tredgold profile generation phase is based on Equations (1)–(14) as outlined in the previous section. The flow-chart in Figure 3 provides a detailed description of the calculation process that can be iterated in order to compute the straight bevel gear tooth profiles with desired performance. The idea is that, having fixed the relative incidence angle between the axes Σ of the gears, the relative number of teeth z_1 and z_2, the module m, and the pitch angles γ_1 and γ_2 can be consequently obtained. Each gear can be treated with the Tredgold method in an iterative manner by considering a certain number of discrete values for the module along the width of the considered gear. It is also possible, for each gear pair element, to obtain a second gear having an orthogonal axis and ensuring conjugate motion. We define them as virtual auxiliary gears. They can have a non-integer number of teeth and cannot be built. However, the gears we are designing will have the same module, and the involutes meshing in a cone perpendicular to their pitch cone do surely mesh.

Table 1 reports an example of gear data for a computed bevel gear pair in order to analyze the contact characteristics of a straight bevel gear pair. Data have been computed with the proposed Tredgold profile generation algorithm by considering a shaft angle of 12°. Hence, the flank data is generated as an ordered set of points, defined by Cartesian coordinates. The cloud-of-points representation is stored in an Excel file and then given as input to a pre-processing Matlab tool that automatically creates the FE mesh of each gear pair element in the gear mesher phase.

The creation of the finite element mesh is customizable by the user. The latter can change the dimension and the number of elements and provide accuracy of the analysis. Biquadratic interpolation techniques, presented in [18], are used in order to accurately build the tooth surfaces starting from a certain number of Cartesian points. Depicted in Figure 4 is an example of the FE mesh creation for the input gear.

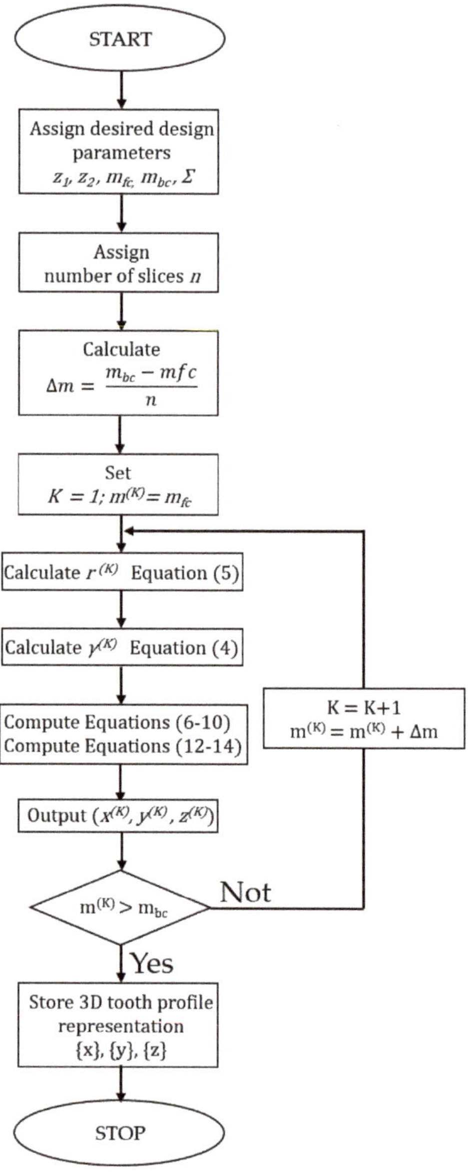

Figure 3. Overview of major steps of the tooth flank generation process.

Table 1. Blank data of the example bevel gear pair.

Gear Data		Pinion	Gear
Number of teeth		14	27
(m_{fc}, m_{bc})	(mm)	(0.9, 1)	(0.9, 1)
n		81	81
Face width	(mm)	9.797	9.797
Mean cone distance	(mm)	93.003	93.003
Pressure angle	(°)	20	20
Pitch radius at midpoint	(mm)	3.318	6.694

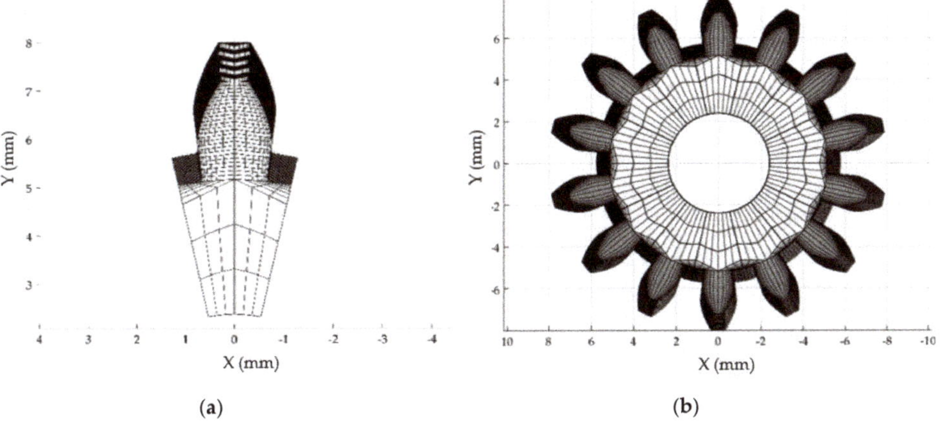

Figure 4. (a) Single tooth finite element mesh; (b) complete gear finite element mesh.

Once the FE mesh is available, the third part of the semi-automated design process consists of numerical tooth contact analysis computation, following the steps below:

- Creation of the FE assembled model.
- Definition of the initial gear pair kinematic configuration.
- Specification of loading campaign and boundary conditions.
- Multiple static nonlinear simulation.
- Post-processing of the tooth contact analysis results.

4. Tooth Contact Analysis

The tooth contact analysis consists of a computerized and numerical procedure that allows identification of the main performance of two mating gears in terms of static transmission error (STE) and distribution of contact pressures along the gear flanks. Since the quality of the motion transmission is strictly dependent on the gear profiles and mainly on the interaction of the mating elements of the gear pair, the TCA plays a key-role in the gear design. Many phenomena occur when two mating gears come into contact. Contact conditions can vary depending on arising misalignments, assembly errors, or micro-modifications. The variability of the contact mechanics leads to nonlinear problems, whose analytical solutions are too difficult to be suitable. Hence, over the last decades of the twentieth century, TCA has been commonly extended with finite element analysis (FEA) [13] or with analytical multibody approaches [10].

In this work, an unloaded tooth contact analysis (UTCA) is computed in order to consider pure geometric conditions for contact detection. On the other hand, results of a loaded tooth contact analysis (LTCA) are presented to provide a more realistic description of the gear pair contact behavior.

The load's application leads to tooth bending deflections. It modifies the overall contact conditions and specifically the contact ratio that is given as the number of teeth, which are instantaneously in contact. If the contact ratio varies, the load sharing and the meshing stiffness vary and, consequently, a vibrational excitation is introduced, as outlined in [19–21].

4.1. FE-Based Tooth Contact Analysis

Due to the occurring nonlinearities, nonlinear FE-based contact simulations have been conducted by means of a commercial software package. The FE model of the gear pair is created using a dedicated preprocessing software tool that allows the creation of discretization of the desired gear flank geometry starting from a cloud-of-points based representation of the three-dimensional gear flanks and the gear blank data. The in-house code of the preprocessing tool allows the user to fully handle the size and the density of the mesh, avoiding problems of convergence and stability of the solutions.

4.1.1. FE Model Creation

The aim of the FE-based tooth contact analysis is to perform the static nonlinear solution in order to investigate the contact zone of the mating gears inside a whole cycle of meshing. Therefore, a FE-based approach requires:

- Development of the finite element mesh of the gear drives.
- Definition of contacting surfaces.
- Establishment of boundary conditions in terms of load and constraints.
- Post-processing of the results.

The FE model of the straight bevel gear pair is obtained by discretizing the gear geometry as a connected set of 3D elements. In this work, six-sided solid elements (HEXA8) have been chosen. Since static nonlinear simulations have significant computational times, the FE model of each gear is characterized by a combination of three FE mesh sections, each with its own mesh density:

- The finer part includes the teeth most likely to enter in contact as the load increases.
- The very coarsely meshed section holds most of the gear that will not be in contact during the contact simulations. It is necessary only to correctly approximate the overall blank stiffness without introducing additional and unneeded degrees of freedom to the model.
- The third part is made up of two small intermediate sections (each on every side) to connect the fine and the coarse mesh parts. Its main task is avoiding discontinuities between the fine and the coarse parts, which could provide bad results or convergence problems for the simulations.

Figure 5 shows the final FE model of the analyzed gear pair.

For the finer part, the number of elements that characterizes the active flanks of the gears is equal to the number of elements of the cloud-of-points representation given as input in the preprocessing phase. This choice has been adopted in order to use the proper input data, while avoiding errors of interpolation techniques during the preprocessing phase.

As it can be seen in Figure 5, 2D rigid body elements (RBE2) were used at the inner part of each gear in order to force the same boundary conditions, defined at the single central node, to all the nodes that lies on the inner cones of the gear pair elements. During the simulation, the gear was kept fully constrained, while the pinion was allowed to rotate about its rotational axis.

The external torque was applied on the pinion's rotational axis. In this work, TCA was performed for five different load cases: 0.01 Nm, 0.1 Nm, 0.5 Nm, and 1 Nm. The studied load cases were chosen considering a maximum load of 0.02 Nm as constrained by the expected operating conditions. The 0.01 Nm load case was defined as the "unloaded case". This was the load that allowed consideration of the pure geometric condition, with no bending tooth deflections, in the mating gears contact analysis. The aim of this work was to propose a suitable design approach for straight bevel gears. The methodology can be used to generate straight bevel gear pairs with any material. In a preliminary prototyping phase, the 3D printed gears represented a proof of the design methodology concept. Moreover, future developments aim at adopting metal powders and sintered element gears. For this reason, the simulations were computed by setting the properties of steel, which is the standard material for geared transmissions. The chosen steel material provided the values of E = 200 GPa (Young modulus) and ν = 0.3 (Poisson ratio) that were assigned to both pinion and gear.

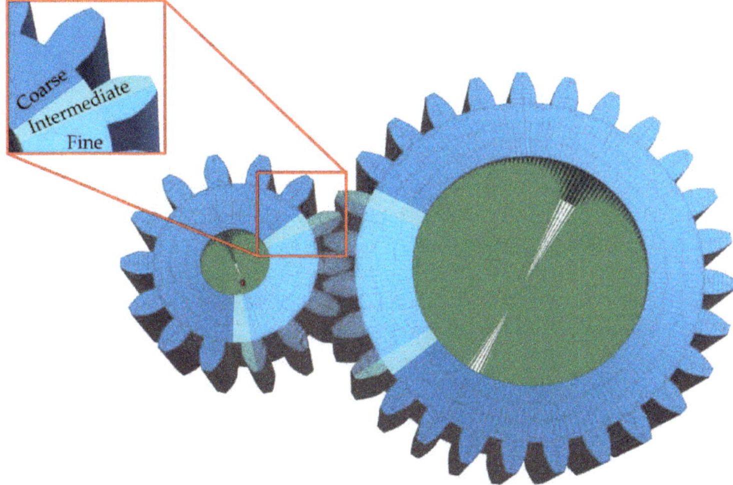

Figure 5. The finite element model of the analyzed straight bevel gear pair.

4.1.2. Kinematic Configuration

Each geared body has to be analyzed in its own kinematic configuration in order to obtain high power transmission performance. Conjugate motion consists of perfect motion transfer that can be maintained for every considered roll angle during a complete meshing cycle. The latter is defined as

$$\phi_{mc} = \frac{2\pi}{z}, \tag{15}$$

where ϕ_{mc} is expressed in radians and z is the number of teeth of the considered gear. In this work, the results refer to the pinion's meshing cycle that is equal to about 0.449 rad. However, toothed bodies have to transmit a certain load during the engagement. Since they are not characterized by infinite stiffness, their profile elastically deforms, giving rise to an unsteady component in the relative angular motion of the mating gears. This component is due to the periodic variation in the stiffness of the gear mesh that can be mainly ascribed to the fluctuation of the contact ratio. The latter is the instantaneous number of teeth in contact during the rotation. Since the meshing stiffness varies, no perfect motion can be transmitted and the mating teeth can no longer be considered conjugate. The condition of non-conjugacy is one of the main sources of dynamic excitation that a gear design tries to minimize at operating conditions.

The static transmission error is an index of the excitation that occurs on the geared systems and of their performances. It considers the difference in rotation between the actual position of the output gear and the position that the gear would occupy in the case of perfect drive.

$$STE = \phi_2 - \frac{z_1}{z_2}\phi_1 = \phi_2 - \tau\phi_1, \tag{16}$$

where ϕ_1 and ϕ_2 represent the rotational angles of the pinion and the driven gear around the correspondent axis of rotation, respectively; z_1 and z_2 are the numbers of teeth and τ is the transmission ratio.

In the presence of pure kinematic motion, the STE should be ideally equal to zero. Equation (16) allows the expression of the STE in μrad as a relative rotation, but it might be usually converted into relative displacement μm by using the pitch radius of the considered gear.

Regarding the starting configuration of the static simulations, the gear pair was oriented in such a way that the mating flanks were in close contact. This operation was not manually done, because the required initial rotational angle of the input gear was determined considering the mean value of the unloaded transmission error (UTE), which is equal to 0.001679 rad. It represents an offset value, since the mean value of STE related to a conjugate gear pair should be null.

4.1.3. Contact Detection Strategy

In order to analyze the contact behavior of the considered gear pair, it was useful to adopt the finite element analysis (FEA). The used contact detection strategy takes place from an FE discretization of the contact interfaces. In this work, the so-called node-to-surface approach was used, which is based on the definition of two kind of contact surfaces: the slave and the master surfaces. According to the considered method, the nodes on the slave surface cannot penetrate the segments of the master surface. The selection of these regions is relevant and results from several considerations about mesh size, surface curvatures, and material. Here, the pinion was chosen to be the slave, while the driven gear was identified as the master.

The node-to-surface method provides a way to find the point $x^{(M)}(\xi,\eta)$ on the master surface $\Sigma^{(M)}$ that minimizes the distance between a given node $x^{(S)}$ of the slave surface $\Sigma^{(S)}$ and the master surface $\Sigma^{(M)}$ [22]. This distance is called gap (g) and it can be expressed as:

$$g = \left[x^{(S)} - x^{(M)}(\xi,\eta) \right] \cdot n^{(M)}, \tag{17}$$

where the gap g is computed as the distance between the position vectors of the slave point $x^{(S)}$ and the master point $x^{(M)}$, projected onto surface normal $n^{(M)}$ that is assessed at the contact point on the master surface. If no friction is considered during the contact, the Hertz–Signorini–Moreau (HSM) condition, presented in Equation (18), can be used as the no-penetration condition in order to impose contact along the normal direction [22].

$$g_N \geq 0, \ \lambda \leq 0, \ \lambda \cdot g_N = 0. \tag{18}$$

While gap is detected, there cannot be a contact force. The latter is expressed in the form of the Lagrange multiplier λ. Conversely, if a contact force is identified, the gap must be null.

In the considered numerical solution, the HSM-condition is replaced by the constraint function presented in Equation (19), where ϵ_N is a small parameter defined by the user [23]:

$$w(g_N, \lambda) = \frac{g_N + \lambda}{2} - \sqrt{\left(\frac{g_N - \lambda}{2} \right)^2 - \epsilon_N}. \tag{19}$$

It is possible to assign a finite compliance to the contact surface. Thus, a user-specified amount of interpenetration between the contacting surfaces is allowed in order to simulate the contact between soft or compliant surfaces. The amount of penetration δ_p is defined as:

$$\delta_p = \epsilon_p \, p_N = \epsilon_p \, \frac{\lambda}{A}, \tag{20}$$

where ϵ_p is the contact surface compliance parameter and p_N is the normal surface contact pressure. Figure 6 shows how the constraint function $w(g_N, \lambda) = 0$ is altered by introducing the contact compliance ϵ_p. The latter allows the emulation of the presence of a unidirectional spring with stiffness constant A/ϵ_p, where A is the contact surface and λ is the contact force that is applied on the contact interface.

Earlier obtained results indicated a maximum contact pressure of about 500 MPa at a load of 1 Nm. Such a high value of the maximum contact pressure and discontinuity of contact patterns induced us to introduce a compliance to the contact surfaces.

Taking into account the size of the gear pair, $\delta = 2.5\ \mu m$ of interpenetration was allowed between both surfaces, yielding a contact compliance value equal to:

$$\epsilon_p = \frac{\delta_p}{p_N} \rightarrow \epsilon_p = \frac{2.5\ \mu m}{500\ MPa} = 0.005 \frac{\mu m}{MPa} = 5 \times 10^{-9} \frac{mm^3}{mN}. \tag{21}$$

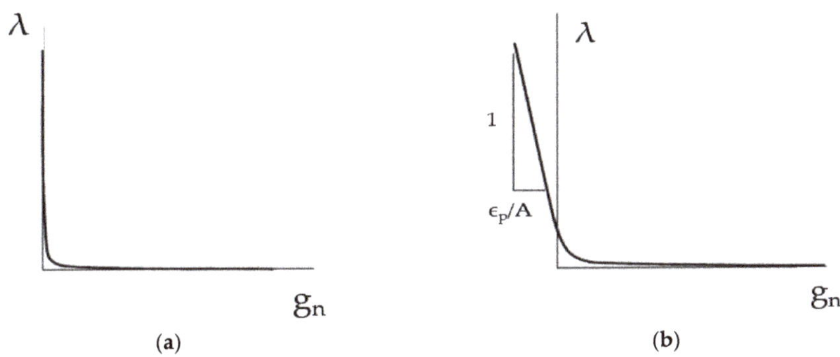

(a)   (b)

Figure 6. (a) Constraint function without contact compliance; (b) constraint function with contact compliance, according to the method described in [23].

4.2. STE Results

An overview of the STE results for the different processed loaded cases is given in Figure 7. The STE analysis allowed us to deduce that in kinematically ideal conditions, the designed straight bevel gears were close to conjugate. This was proved by the flat trend of the unloaded transmission error (UTE). Increasing the applied load, the effect of tooth bending could be observed. Indeed, when bending deformations occurred, the number of teeth in contact changed, as did the mesh stiffness and the peak-to-peak (p-p) of the STE curves, as reported in Table 2.

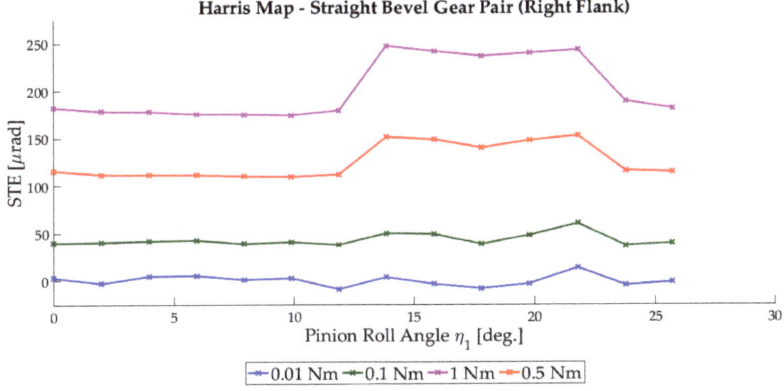

Figure 7. Static transmission error (STE) curves of the bevel gear pair under different loading conditions.

Table 2. Calculated peak-to-peak values under different loading conditions.

Peak-to-Peak STE (µrad)			
0.01 Nm	0.1 Nm	0.5 Nm	1 Nm
22.67	23.65	43.35	73.10

Figure 8 shows the STE curve for 1 Nm. Two different regions are highlighted: the first part is characterized by small values of STE, because in those configurations two pairs of teeth were in contact and so the mesh stiffness was bigger; the second part presents higher values of STE, which means that the mesh stiffness decreased. A decrease of mesh stiffness was due to the reduction of the number of teeth in contact, from two pairs to only one pair.

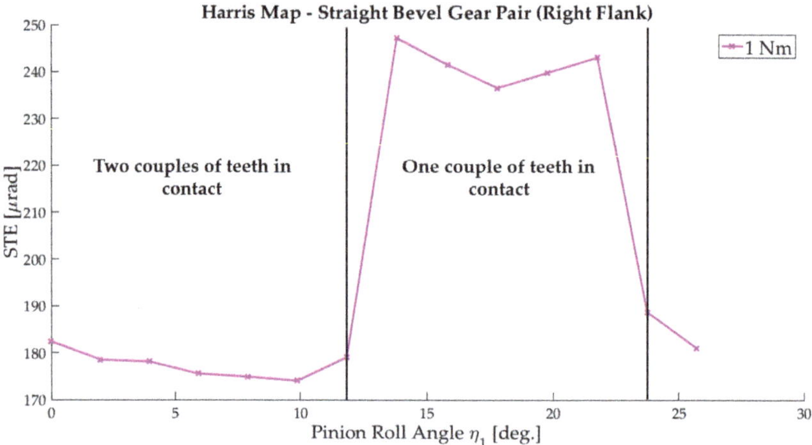

Figure 8. STE curve for 1 Nm.

A parameter that quantifies the instantaneous number of teeth pairs in contact is the contact ratio m_c. For straight bevel gears, it is possible to adopt the spur gears' expression, as reported in [12], considering the back-cone equivalent gears:

$$m_c = \frac{g_a}{r_p} = 1.53, \tag{22}$$

where g_a is the length of path contact and r_p is the base pitch. The next lower integer of m_c indicates the average number of pairs of teeth in contact. Thus, a contact ratio of $m_c = 1.53$ means that there is always at least one pair of teeth in contact and there are two pairs of teeth in contact 53% of the meshing cycle.

The same result could be obtained focusing the attention on Figure 8, because the STE curve along the meshing cycle can be divided in two parts that have similar lengths.

A contact pressure analysis was considered in order to give a whole TCA scenario. The mesh stiffness variation up to 1 Nm was confirmed, as seen in Figure 9, where the transition between two pairs (a) and one pair (b) of teeth in contact is shown. The contact pattern consisted of a line as expected for straight bevel gears. Similarly, the maximum contact pressure decreased when two pairs of teeth were in contact due to the load sharing. It is also important to note that increasing the contact ratio implied increasing load sharing, reducing the maximum contact pressures, and improving motion transmission smoothness.

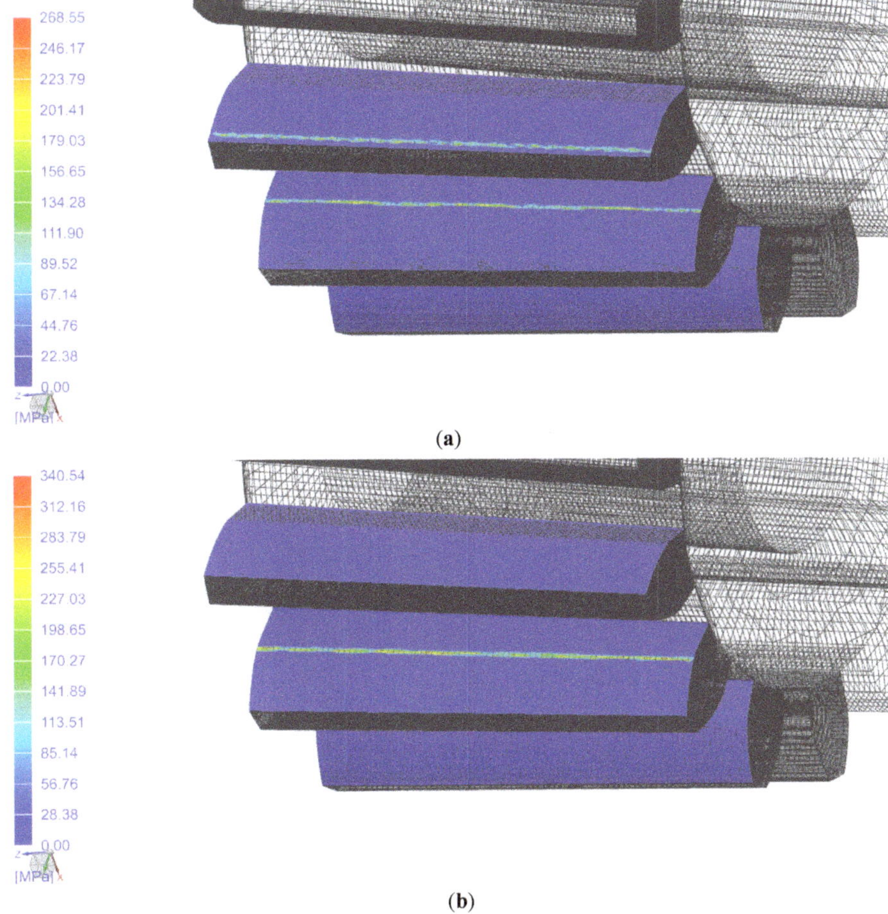

Figure 9. Contact patterns of two consecutive steps along the mesh cycle: (**a**) when two pairs of teeth are in contact; (**b**) when one pair of teeth is in contact.

5. Discussion

The obtained numerical results, depicted in Figure 7, allow us to underline the effectiveness of the proposed design algorithm for the considered case of study. Namely, it is considered a biomedical application, where the gear train will operate with very slow and accurate motions. It is worth noting that the main source of dynamic excitation of the geared transmission systems comes from the variability of the meshing stiffness at operating conditions. Consequently, small variations of STE, and thus small values of peak-to-peak (p-p), represent a desired outcome of the design procedure for the considered gear pair.

As reported in the literature [19], reasonable levels of TE peak-to-peak depend on the specific application of the geared system. Two gear pairs that differ in size or in shape can be compared in terms of noise and vibration performance if their STE are expressed in μm [19]. Large values of STE p-p would be permissible on large, slow-speed geared machinery, where the gear noise usually does not represent a significant problem. At the ultra-precision end, a TE of 1 μm p-p could be considered as extremely good.

Focusing the attention on Table 2, the peak-to-peak TE values increase with the applied load, because of the increasing of tooth bending. The highest value is about 73.10 μrad. If the latter is multiplied by the pinion's pitch radius, as reported in Table 1, the maximum peak-to-peak value is obtained at about 0.24 μm. Hence, this value confirms a good vibrational performance and the designed gear pair can be considered as suitable for the specific application for the transmission system of the EasyLap robotic system, where high precision and small footprint are prescribed.

Future developments of the presented algorithm could include:

- Adding the STE as an optimization variable in the process of calculation of the radii and the angles, while the objective function could be represented by the minimum allowable encumbrance.
- The possibility of introducing micro-geometry modifications of the tooth profile in order to minimize the STE peak-to-peak values, as reported, for example, in [21].

A 3D printed prototype is shown in Figure 10. This prototype has been successfully mounted at the wrist of the EasyLap operating arm. Figure 11 shows the sterilizable component of the adaptor for standard laparoscopic instruments using the designed couple of small size bevel gears.

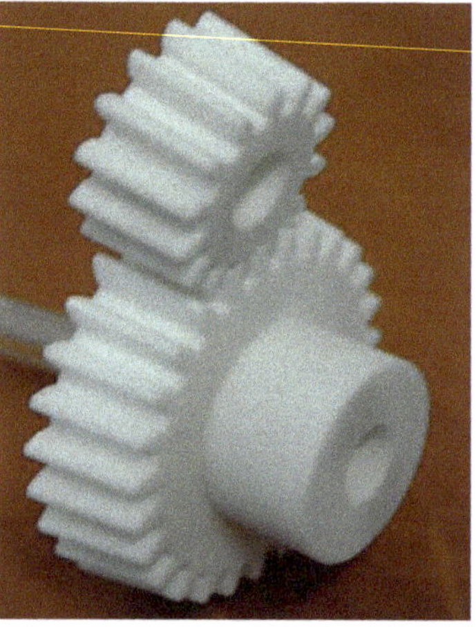

Figure 10. 3D printed prototypes of the bevel gear pair elements.

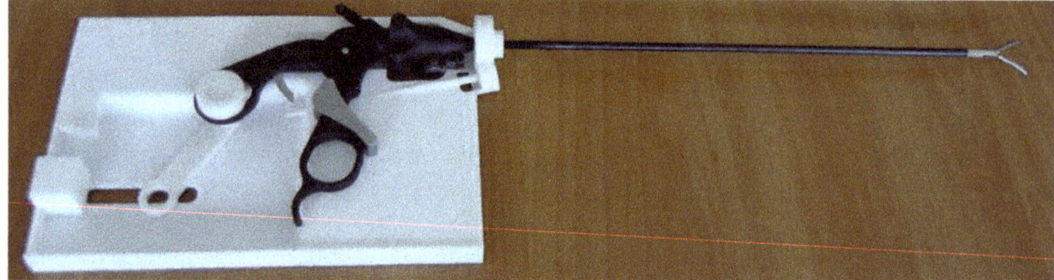

Figure 11. 3D printed adaptor for traditional surgical instruments using this gear pail.

6. Conclusions

This paper has proposed a semi-automated design algorithm for computing straight bevel gear involute profiles by taking into account a 3D printing manufacturing process. The proposed formulation relies on Tredgold approximation, and is suitable for any desired number of teeth and relative axes inclination angles. A specific case study was carried out to design a miniaturized bevel gear pair in order to minimize the encumbrance for a new laparoscopic robotic system, EasyLap. A meshing analysis and a finite element-based tooth contact analysis were carried out in order to evaluate the behavior of the geared transmission under a set of different loads that did not overcome the applied working conditions. The numerical simulations demonstrated that the proposed non-traditional design approach provides conjugate tooth profiles of the contacting surfaces as well as their suitable vibrational performances. The latter are properly used to guarantee high precision for biomedical applications. Finally, a prototype was also built and successfully installed on an EasyLap robotic system, proving engineering feasibility and effectiveness of the proposed design procedure. Future developments will extend the proposed algorithm, including an STE-based optimization in the calculation of the radii and angles.

Author Contributions: Conceptualization, A.A., G.F., P.F.G., and G.D.; Investigation, G.D.; Methodology, A.A. and G.D.; Supervision, D.M., G.C., and G.D.; Visualization, A.A.; Writing— original draft, A.A.; Writing— review and editing, D.M., G.C., and G.D..

Acknowledgments: This work was performed under project EasyLap, POR Calabria 2014-2020 Fesr-Fse, CUP J28C17000130006. The authors wish to acknowledge the help of Gianluca La Greca and Basilio Sinopoli for preparing the CAD model and printing the prototypes of the bevel gears. The authors gratefully acknowledge Siemens Industry Software NV (Belgium) for the valuable technical support.

References

1. Kalk, H.; Brühl, W. *Leitfaden der Laparoskopie und Gastroskopie*; Georg Thieme Verlag: Stuttgart, Germany, 1951. [CrossRef]

2. Berci, G.; Cuschieri, A. *Practical Laparoscopy*; Bailliere Tindall: London, UK, 1986.

3. Diaz-Arrastia, C.; Jurnalov, C.; Gomez, G.; Townsend, C. Laparoscopic hysterectomy using a computer-enhanced surgical robot. *Surg. Endosc. Other Interv. Tech.* **2002**, *16*, 1271–1273. [CrossRef] [PubMed]

4. Advincula, A.; Song, A.; Burke, W.; Reynolds, R. Preliminary Experience with Robot-Assisted Laparoscopic Myomectomy. *J. Am. Assoc. Gynecol. Laparosc.* **2004**, *11*, 511–518. [CrossRef]

5. Di Marco, D.; Chow, G.; Gettman, M.; Elliott, D. Robotic-assisted laparoscopic sacrocolpopexy for treatment of vaginal vault prolapse. *Urology* **2004**, *63*, 373–376. [CrossRef] [PubMed]

6. Sackier, J.M.; Wang, Y. Robotically assisted laparoscopic surgery. From concept to development. *Surg. Endosc.* **1994**, *8*, 63–66. [CrossRef] [PubMed]

7. Reichenspurner, H.; Damiano, R.J.; Mack, M.; Boehm, D.H.; Gulbins, H.; Detter, C.; Meiser, B.; Ellgass, R.; Reichart, B. Use of the voice-controlled and computer-assisted surgical system ZEUS for endoscopic coronary artery bypass grafting. *J. Thorac. Cardiovasc. Surg.* **1999**, *118*, 11–16. [CrossRef]

8. Babbar, P.; Hemal, A.K. Robot-assisted urologic surgery in 2010—Advancements and future outlook. *Urol. Ann.* **2011**, *3*, 1–7. [PubMed]

9. Fragomeni, G.; Gatti, G.; Greco, P.F.; Nudo, P.; Perrelli, M.; Sinopoli, B.; Rizzuto, A.; Danieli, G. EasyLap—New robotic system for single and multiple access laparoscopy using almost only traditional laparoscopic instrumentation. In Proceedings of the International Symposium of Mechanism and Machine Science, Baku, Azerbaijan, 11–14 September 2017.

10. Vivet, M.; Mundo, D.; Tamarozzi, T.; Desmet, W. An analytical model for accurate and numerically efficient tooth contact analysis under load, applied to face-milled spiral bevel gears. *Mech. Mach. Theory* **2018**, *130*, 137–156. [CrossRef]

11. Dooner, D.B.; Vivet, M.; Mundo, D. Deproximating Tredgold's Approximation. *Mech. Mach. Theory* **2016**, *102*, 36–54. [CrossRef]

12. Uicker, J.J.; Pennock, G.R.; Shigley, J.E. *Theory of Machines and Mechanisms*; Oxford University Press: New York, NY, USA, 2017.

13. Litvin, F.L.; Fuentes, A. *Gear Geometry and Applied Theory*; Cambridge University Press: New York, NY, USA, 2004.

14. Gupta, K.; Jain, N.K.; Laubscher, R. *Advanced Gear Manufacturing and Finishing, Classical and Modern Processes*; Academic Press: London, UK, 2017.

15. Ozel, C.; Inan, A.; Ozler, L. An investigation on Manufacturing of the Straight Bevel Gear using End Mill by CNC Milling machine. *J. Manuf. Sci. Eng.* **2005**, *127*, 503–511. [CrossRef]

16. Shih, Y.-P. Mathematical model for Face-Hobbed Straight Bevel Gears. *J. Mech. Des.* **2012**, *134*. [CrossRef]
17. Kapelevich, A. Gear design: Breaking the status quo. *Mach. Des.* **2007**, *79*, 89–93.
18. Vivet, M.; Heirman, G.H.K.; Tamarozzi, T.; Desmet, W.; Mundo, D. An Ease-off Based Methodology for Contact Detection and Penetration Calculation. In Proceedings of the International Conference on Power Transmissions, Chongqing, China, 27–30 October 2016.
19. Smith, J.D. *Gear Noise and Vibration*; Marcel Dekker Inc.: New York, NY, USA, 2003.
20. Elkholy, A.H.; Elsharkawy, A.A.; Yigit, A.S. Effect of Meshing Tooth Stiffness and Manufacturing Error on the Analysis of Straight Bevel Gears. *J. Mech. Struct. Mach.* **1998**, *26*, 41–61. [CrossRef]
21. Korta, J.A.; Mundo, D. Multi-objective micro-geometry optimization of gear tooth supported by response surface methodology. *Mech. Mach. Theory* **2017**, *109*, 278–295. [CrossRef]
22. Wriggers, P. *Computational Contact Mechanics*; John Wiley and Sons: Chichester, UK, 2002.
23. *Advanced Nonlinear Solution—Theory and Modeling Guide*; NX Nastran 10; Siemens Product Lifecycle Management Software Inc.: Plano, TX, USA, 2014.

Article

The Design of a New Manual Wheelchair for Sport

Giuseppe Quaglia *, Elvio Bonisoli and Paride Cavallone

Department of Mechanical and Aerospace Engineering, Politecnico di Torino, 10129 Torino, Italy;
elvio.bonisoli@polito.it (E.B.); paride.cavallone@polito.it (P.C.)
* Correspondence: giuseppe.quaglia@polito.it

Received: 28 February 2019; Accepted: 6 May 2019; Published: 9 May 2019

Abstract: In this paper, an innovative system of propulsion inspired by a rowing gesture for manual wheelchairs is shown. The innovative system of propulsion, named Handwheelchair.q, can be applied to wheelchairs employed in everyday life and to sports wheelchairs for speed races, such as Handbike and Wheelchair racing. The general features of the innovative system of propulsion and the functional designs of the different solutions are described in detail. In addition, the design of the mechanism for the transmission of motion, employed in a second prototype, Handwheelchair.q02, is presented and analysed. Finally, the dynamic model of the Handwheelchair.q has been developed in order to obtain important results for the executive design of Handwheelchair.q.

Keywords: Handwheelchair.q; Disabled sport; Manual wheelchair

---

1. Introduction

Many studies [1–3] assert that wheelchair users suffer from upper limb injuries more frequently than the rest of the population. Since the handrim is the world's most used manual propulsion system [4], the main causes of upper limb pain depend on the motion of the handrim system that is a pushing movement. Outdoor and indoor motor activities are extremely important for the well-being of everyone. For disabled people, motor activities [5] are also an important tool in rehabilitation from a physical and psychological point of view. Even though motor activities for disabled people are extremely important, they are not easy to practice. First of all, specific sports require specific wheelchairs [6–12]; secondly, the areas where the disabled can practice sport have to be properly equipped; finally, the disabled are not independent for the transportation of the wheelchair for a specific sport.

The above reasons motivated us to develop a wheelchair with an innovative system of propulsion, which not only allows for the practice of outdoor activities autonomously, but also eliminates the need of a specific wheelchair for sport.

As shown in [13], during a propulsion cycle with the handrim system, the glenohumeral contact force has high peaks. This is one of the factors that can increase the risk of injuries to the shoulders. In addition, in the Conclusions section of [14] "The latter makes tangential force propulsion not only less efficient, but also more straining for the shoulder". Different articles compare the handrim system with alternative systems of propulsion, such as a lever system [15] and a handcycle [13].

Another approach considers the use of a rowing gesture in order to obtain the propulsion. Furthermore, the rowing stroke [16] is divided into four phases and different articles have studied the efficiency of the rowing gesture [17]. The rowing stroke is a complex movement; the muscles of the legs, trunk, back, shoulders and arms are involved. The rowing stroke can be performed by disabled athletes with different levels of spinal injuries [18]. Disabled people can perform this movement depending on the height of the spinal injuries.

In the last two years, the first prototype, named Handwheelchiar.q01 [19], was released. The first prototype tested the functioning of the innovative system of propulsion and highlighted some critical issues.

In addition to the Handwheelchair.q01, the authors of this paper have wide experience with means of mobility for disabled people [20,21].

2. Rowing Gesture: General Features

The innovative system of propulsion is inspired by the rowing gesture. One of the more important characteristics of the rowing gesture is that the drive phase is obtained using a pulling movement of the arm, instead of a pushing one, typical of handrim and lever systems. The pulling movement could be a good alternative compared to the pushing movement. In this work, a cable solution was introduced in order to realise the rowing motion as a system of propulsion. The cable solution allows users to optimise their motion based on their own individual physical characteristics.

The rowing gesture is composed of two phases. During the traction phase, the user provides power, while in the recovery phase the user goes back to the initial position.

During the traction phase shown in Figure 1, the user pulls two cables by two handles wrapped around two pulleys. Each pulley transmits the torque and rotation to the wheel by a unidirectional mechanism, named a ratchet system. A power spring, which connects the pulley to the chassis of the wheelchair, is loaded. In this phase, the angular speed of the pulley and the wheel are equal.

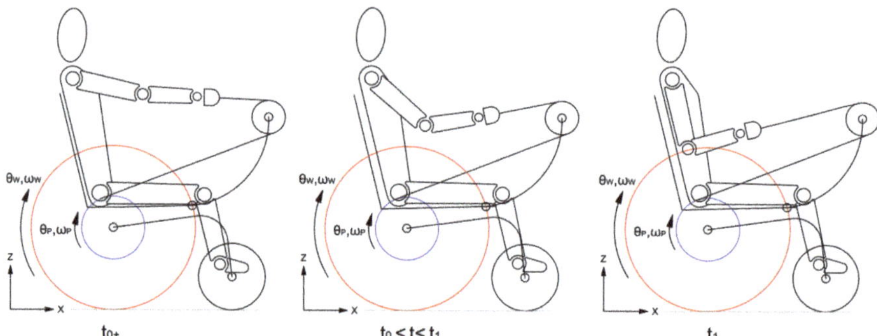

Figure 1. Rowing gesture during the traction phase.

During the recovery phase shown in Figure 2, the user stops pulling and the power spring, previously loaded, rotates the pulleys in the opposite direction ($\omega_w > 0$, $\omega_P < 0$). The cables are wrapped around each pulley and the user can start another traction phase.

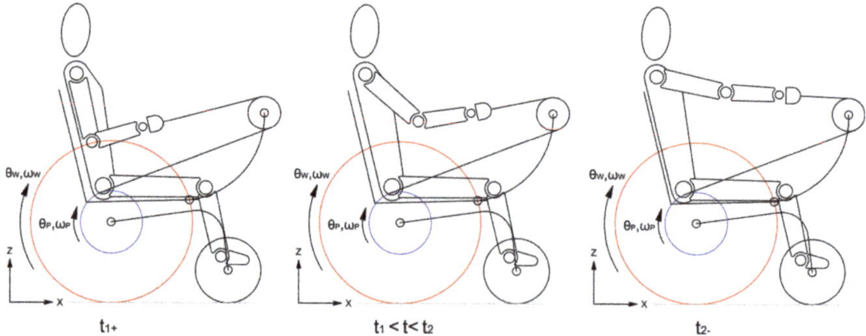

Figure 2. Rowing gesture during the recovery phase.

In Figure 3, the angular position, the angular speed of the pulley and the wheel are shown with the three assumptions:

- The angular speeds are constant;
- The time of the traction phase T_P is higher than the time of the recovery phase T_R;
- Pure rolling motion.

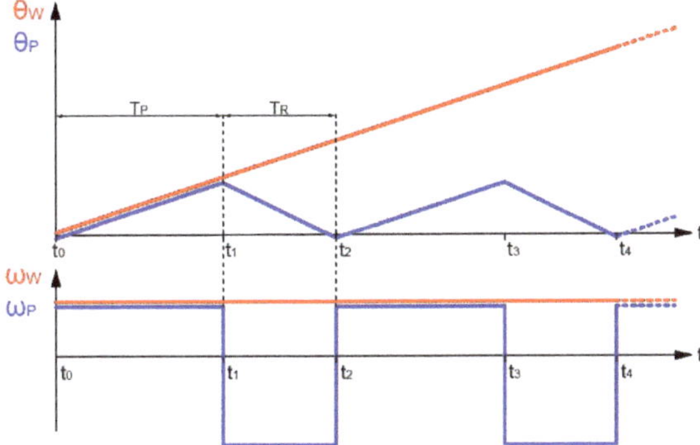

Figure 3. Angular speed and angular position of the pulley and the wheel.

In Section 4, the mechanism of transmission of motion will be accurately described.

3. Application of the Innovative System in Sport

The innovative system of propulsion can be employed by different means of manual mobility for sport: Handbike and Wheelchair racing.

The innovative system can be installed on the Handwheelchair.q racing wheelchair, as shown in Figure 4, with a few modifications of standard racing wheelchairs. The push-rims have been removed, while a pair of traction pulleys and a pair of return pulleys have been placed. The Handwheelchair.q racing wheelchair with the innovative system of propulsion keeps the same characteristics of a racing wheelchair: two rear traction wheels and a front steering wheel.

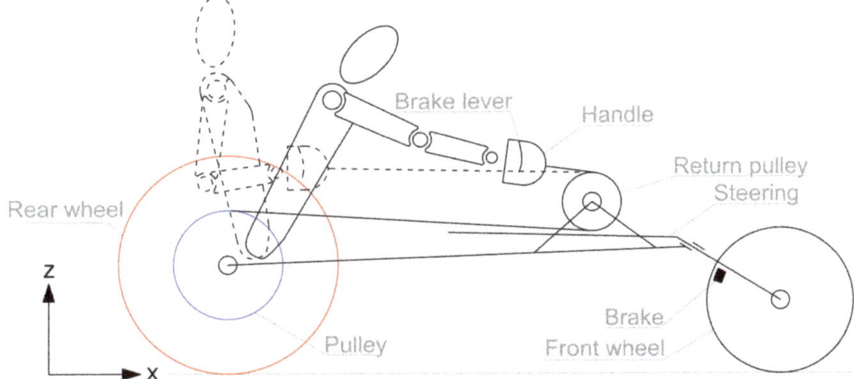

Figure 4. Innovative system applied to the racing wheelchair.

The innovative system can be employed on the Handbike. Figure 5a shows the functional design of Handbike.q for categories H1, H2, H3 and H4, namely disabled with spinal cord injuries and Figure 5b for categories H5, namely disabled with amputated limbs. A pair of pulleys has been seated, one on each side, in order to balance the fork transversely. A return pulley has been located in order to optimise the athlete's motion. The Handbike.q with the innovative system of propulsion retains the same characteristics of the Handbike: a traction steering front wheel.

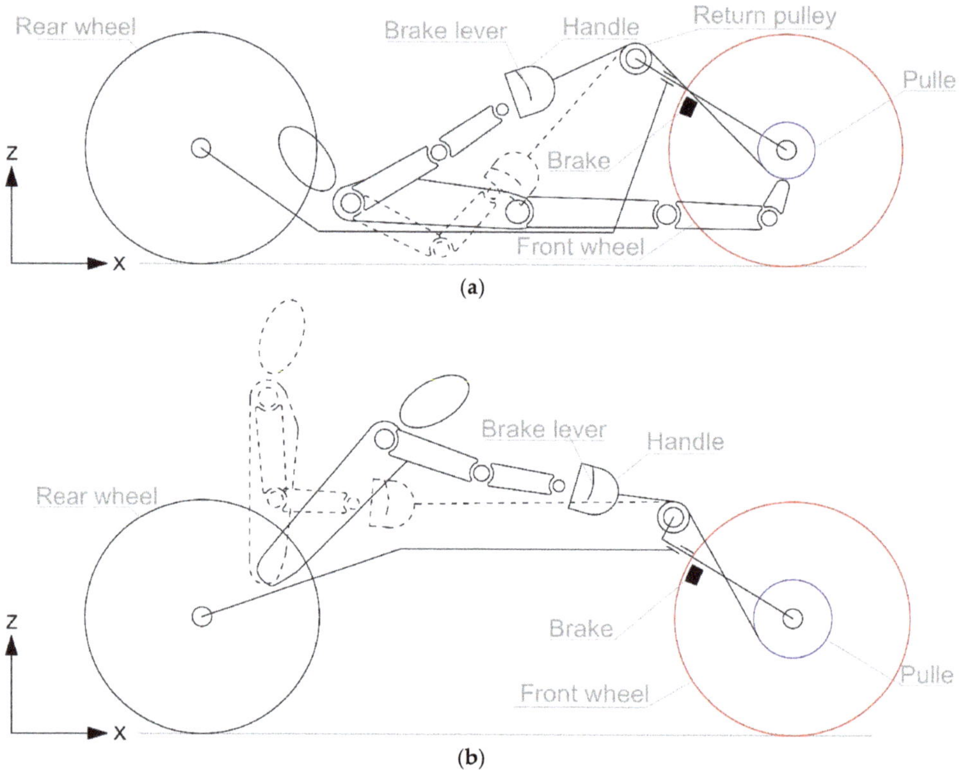

Figure 5. (a) Innovative system applied to the recumbent Handbike; (b) Innovative system applied to the Handbike.

4. Handwheelchair.q02

The Handwheelchair.q02 is a wheelchair for everyday life. The main goals of Handwheelchair.q02 are:

- To facilitate and extend outdoor movement;
- To be firm and maneuverable for indoor spaces; and
- To practice motor activity independently and in such areas as a park and a cycle path.

The Hanwheelchair.q02 has two configurations. The first configuration, shown in Figure 6a, is employed for indoor spaces. In this configuration, the wheelchair is used as a common wheelchair, where the propulsion is obtained by the handrim. The foldable links, left and right, that support a pair of return pulleys, are folded in order to minimise the visual impact and to allow for sitting on the wheelchair. The second configuration, shown in Figure 6b, is employed outdoors to facilitate movement, to extend accessible places and to practice motor activities. The user rotates the foldable link, left and right, around the joint C in order to obtain the innovative configuration.

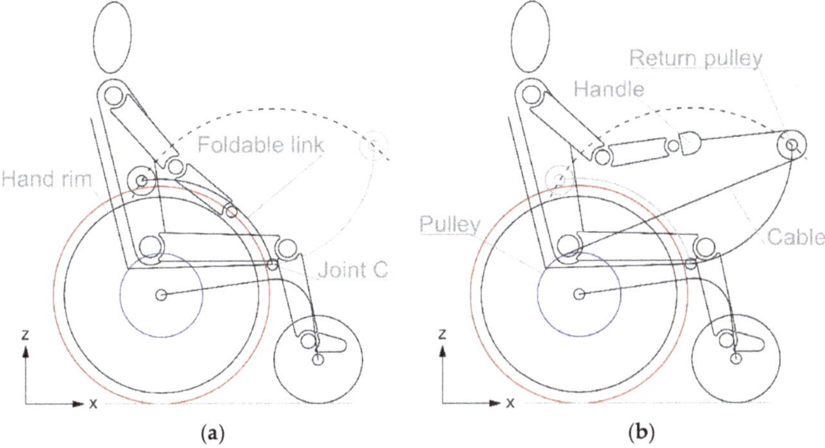

Figure 6. (**a**) Handwheelchair.q02 indoor configuration; (**b**) Handwheelchair.q02 outdoor configuration.

Figure 7 show the components that enable the use of the wheelchair in the innovative configuration. The red component is the hub of the wheel and the blue one is the pulley around which the cable is wrapped. The green one is the power spring that connects the pulley with the chassis of the wheelchair, in grey. In the first configuration, the mechanism of transmission of motion allows the user to use the wheelchair as a common wheelchair. In fact, the hub of the wheel rotates around the shaft as in the classic wheelchair, as sketched in Figure 8a.

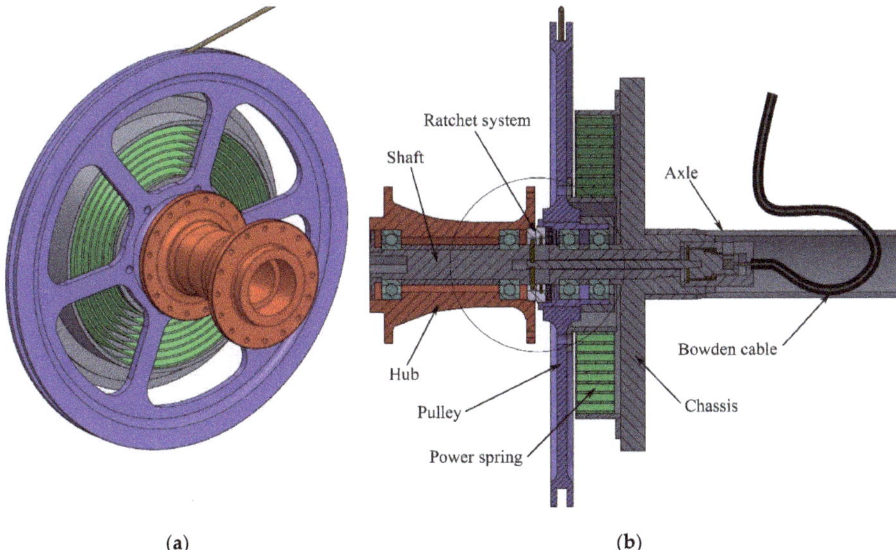

Figure 7. (**a**) Components of the innovative system of propulsion; (**b**) Section of the mechanism.

When the user takes the handles from their site, a mechanism keeps in contact the two front ratchets, as visible in Figure 8b.

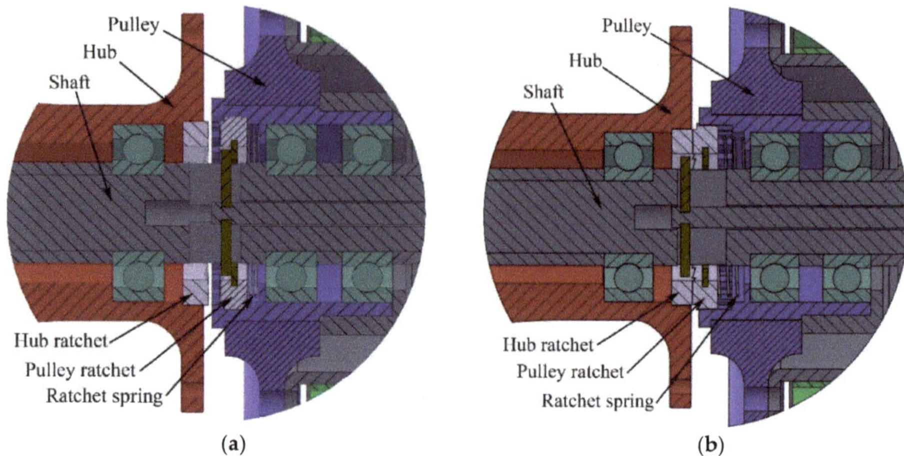

(**a**)   (**b**)

Figure 8. Particulars of the mechanism in the (**a**) Classic configuration; (**b**) Innovative configuration.

One ratchet is integrated with the pulley and the other one is mounted into the hub of the wheel. In this configuration, the propulsion is obtained by the rowing stroke previously described. During the traction phase, the user pulls the cables with a force Fu the cables are wrapped around a pair pulleys of radius rp. The user transmits a torque to the wheel through the ratchet system. During the traction phase, the user loads a power spring that connects the pulley with the chassis of the wheelchair. In Figure 9, the free-body diagrams during the traction phase of the pulley and of the hub are reported and in Tables 1 and 2 the description of the torques and the reference systems are described.

Table 1. Description of the torques in the free-body diagrams.

Torque	Description
$Fu*rp$	User force multiplied the radius of the pulley
Ceo	Preload of the power spring
$Ke*\theta_P$	Stiffness of the power spring multiplied by the angular position of the pulley
$Ip*\dot{\omega}_P$	Moment of inertia of the pulley
$C1$	Torque transmitted by the ratchets
$I_W*\dot{\omega}_W$	Equivalent inertia on the wheel given by the user, the wheel and the wheelchair multiplied by the angular acceleration of the wheel
$Cf1$	Friction torque given by rolling resistance and the aerodynamic resistance
$Cf2$	Friction torque given by the friction ratchets during the recovery phase

Table 2. Reference systems.

Reference System	Description
$\theta_P,\ \omega_P,\ \dot{\omega}_P$	Angular position, angular speed and angular acceleration of the pulley
$\theta_W,\ \omega_W,\ \dot{\omega}_W$	Angular position, angular speed and angular acceleration of the wheel

During the traction phase, the angular speed of the pulley is the same as the angular speed of the wheel and the equation can be written as follows:

$$\omega_P = \omega_W > 0. \tag{1}$$

The torque equilibrium of the pulley results:

$$Fu * rp = I_P * \dot{\omega}_P + Ceo + Ke * \theta_P + C1. \tag{2}$$

While the torque equilibrium of the hub is:

$$C1 = Cf1 + I_W * \dot{\omega}_W. \tag{3}$$

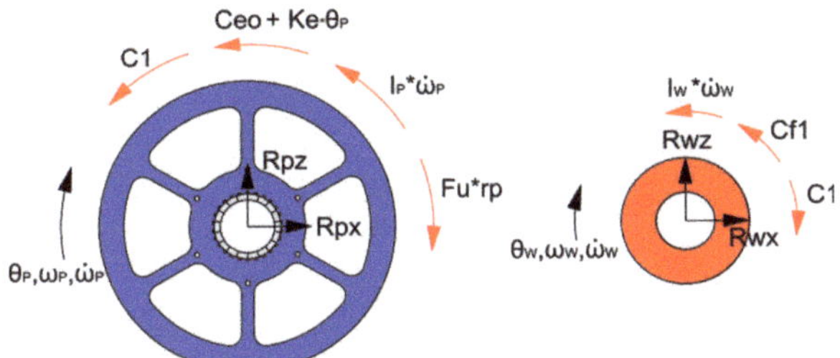

Figure 9. Free-body diagram during the traction phase.

During the recovery phase the user stops pulling, and we assume that Fu ~ 0, Figure 10. The power spring, previously loaded, has to generate a torque Ceo + Ke*θP to rotate the pulley in the opposite direction in order to rewind the cable around the pulley in a specific time T_R. The recovery time T_R is the parameter of the project that defines the torque of the power spring.

During the recovery phase, the angular velocities are:

$$\omega_P < 0; \ \omega_W > 0. \tag{4}$$

The torque equilibrium of the pulley is:

$$I_P * \dot{\omega}_P + Ceo + Ke * \theta_P = Cf2. \tag{5}$$

The torque equilibrium of the hub results:

$$Cf1 + Cf2 + I_W * \dot{\omega}_W = 0. \tag{6}$$

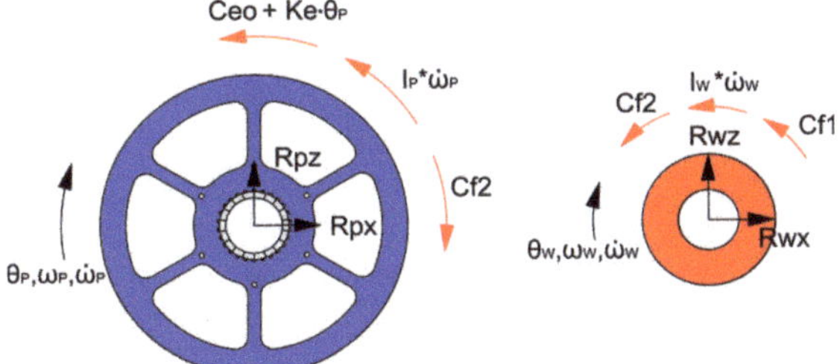

Figure 10. Free-body diagram during the recovery phase.

5. Dynamic Model of Handwheelchair.q

In this section, according to the previous paragraph, a simplified dynamic model, Figures 11 and 12, and the description of the free-body diagram, Tables 3 and 4, of Handwheelchair.q are presented, with the following hypothesis and assumptions:

-	All dynamic characteristics are concentrated on the wheel 1: N2 = 0, T2 = 0;

- Even if the rolling-resistance coefficient depends on the speed [22], the rolling-resistance coefficient u1 can be assumed to be constant for low speed;
- The inertia of the wheels and the pulleys is equal to zero: $I_{W1} = 0$; $I_{W2} = 0$, $I_P = 0$; $I_{RP} = 0$;
- The torque of the power spring Ce is constant;
- Fa is the aerodynamic force, modelled by [22], $Fa = k_a \dot{x}^2$ applied on the Center of Pressure; and
- The Center of Pressure coincides with the Center of Mass CM.

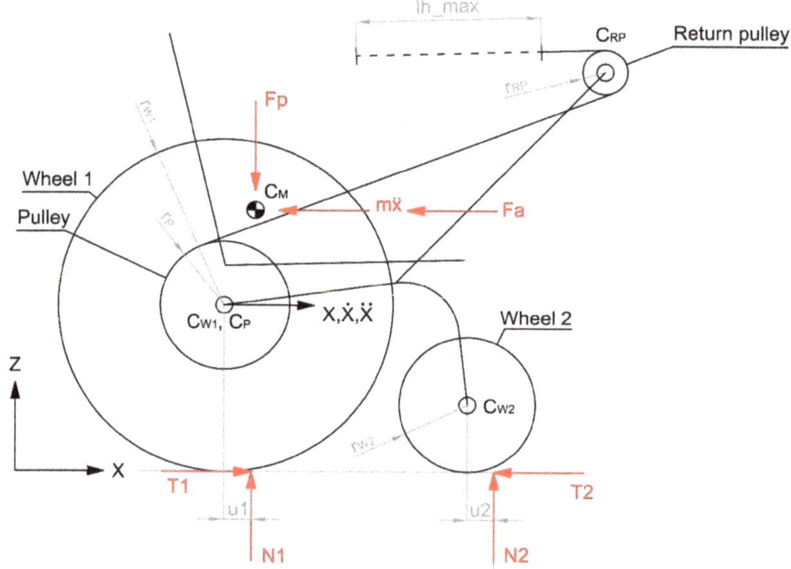

Figure 11. Handwheelchair.q: Free-body diagram.

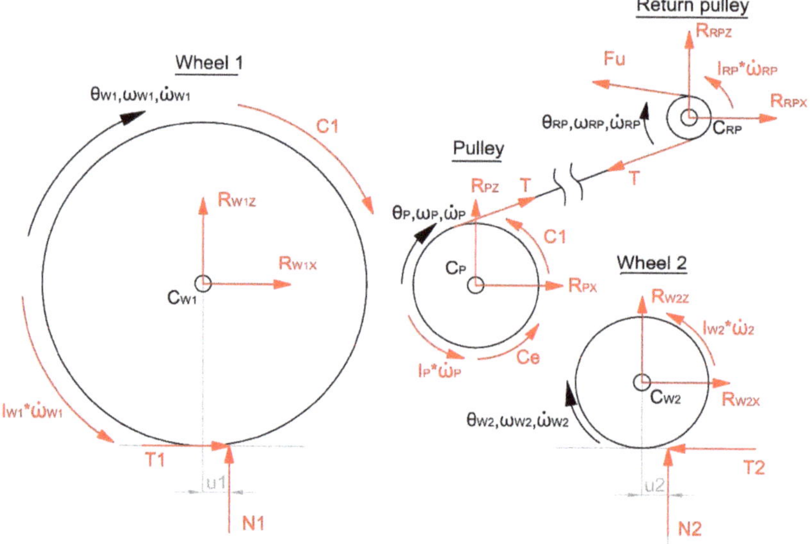

Figure 12. Free-body diagram of wheel 1, wheel 2, the pulley and the return pulley.

Table 3. Description of the torques and force of the dynamic model.

Torque/Force	Description
Fp	Weight force of the user and wheelchair
$m\ddot{x}$	Inertial force of the user and wheelchair
Fa	Aerodynamic force
N1	Ground reaction force of the wheel 1 along the z axes
T1	Ground reaction force of the wheel 1 along the x axes
N2	Ground reaction force of the wheel 2 along the z axes
T2	Ground reaction force of the wheel 2 along the z axes
T	Friction torque given by the friction ratchets during the recovery phase
R_{W1X}	Joint reaction force of the wheel 1 along x axes
R_{W1z}	Joint reaction force of the wheel 1 along z axes
R_{W2X}	Joint reaction force of the wheel 2 along x axes
R_{W2X}	Joint reaction force of the wheel 2 along z axes
R_{PX}	Joint reaction force of the pulley along x axes
R_{PZ}	Joint reaction force of the pulley along z axes
R_{RPX}	Joint reaction force of the return pulley along x axes
R_{RPZ}	Joint reaction force of the return pulley along z axes
$I_{W1} * \dot{\omega}_{W1}$	Moment of inertia of the wheel 1
$I_{W2} * \dot{\omega}_{W2}$	Moment of inertia of the wheel 2
$I_{P} * \dot{\omega}_{P}$	Moment of inertia of the pulley
$I_{RP} * \dot{\omega}_{RP}$	Moment of inertia of the return pulley

Table 4. Reference system's dynamic model.

Reference System	Description
$x, \dot{x}, \ddot{x}$	Position, speed, and acceleration of the wheelchair
$\theta_P, \omega_P, \dot{\omega}_P$	Angular position, angular speed, and angular acceleration of the pulley
$\theta_{RP}, \omega_{RP}, \dot{\omega}_{RP}$	Angular position, angular speed, and angular acceleration of the return pulley
$\theta_{W1}, \omega_{W1}, \dot{\omega}_{W1}$	Angular position, angular speed, and angular acceleration of the wheel 1
$\theta_{W2}, \omega_{W2}, \dot{\omega}_{W2}$	Angular position, angular speed, and angular acceleration of the wheel 2

The user's force is modelled as follows:

- The time of the traction phase is Tp and, according to Figure 13, is evaluated as:

$$x(t') - x(t0) = lh\_max * \tau \tag{7}$$

$$t' - t0 = Tp \tag{8}$$

where, lh is the stroke length of the gesture and $\tau = 3$ is the transmission ratio defined by:

$$x = lh * \frac{rw1}{rp} = lh * \tau. \tag{9}$$

The Figure 14, shows the traction phase during the indoor test of the first prototype Handwheelchair.q01.

- According to [12,13], during the tests the propulsion power was between 15 and 55 W. The average propulsion power for a cycle can be assumed to be between 20 and 25 W;
- The user's force Fu has been modelled with a polynomial 2-3-4 in order to define an approximate model of the "Handle force Fh" [23], as shown in Figure 15;
- The scale of force Fu is determined in order to obtain the average power of 20–25 W;
- The recovery time Tr is constant;
- The user's force during the recovery phase is zero.

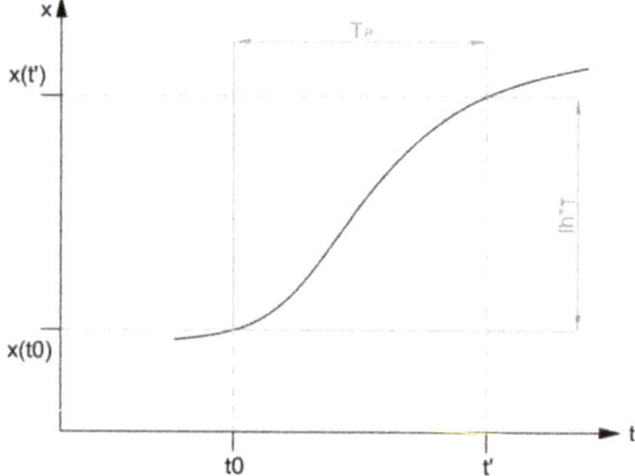

Figure 13. Time of the traction phase.

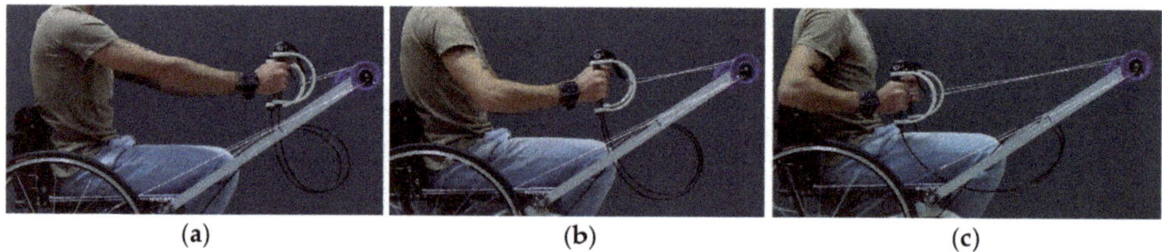

(a) (b) (c)

Figure 14. Innovative gesture during the indoor test of the first prototype Handwheelchair.q01. (**a**) t_0, lh = 0; (**b**) $t_0 < t < t_1$, $0 < $ lh $<$ lh_max; (**c**) t_1, lh = lh_max.

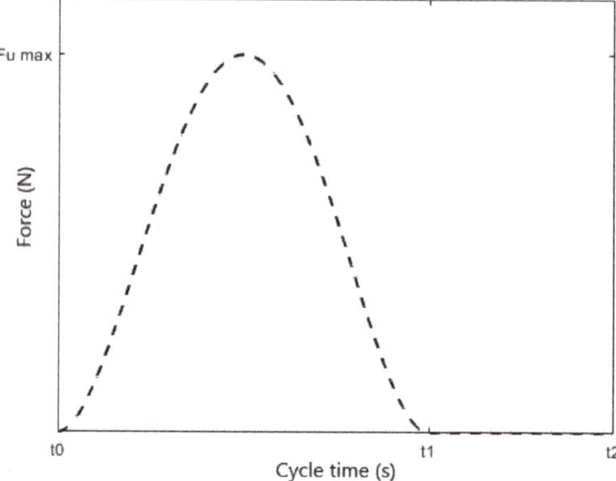

Figure 15. User's force model.

According to the free-body diagram previously reported, and with the abovementioned hypothesis and assumptions:

The equilibrium of the wheelchair along the axis x results

$$T1 = m\ddot{x} + Fa = m\ddot{x} + k_a\dot{x}^2. \tag{10}$$

The equilibrium of the wheelchair along the axis z is

$$N1 = Fp = mg. \tag{11}$$

The rotation equilibrium of wheel 1 around the joint C_{w1} results

$$T1 * r_{W1} + N1 * u_1 = C1. \tag{12}$$

The rotation equilibrium of the pulley around the joint C_P is

$$T * r_P = C1 + Ce. \tag{13}$$

The rotation equilibrium of the return pulley around the joint C_{RP} is

$$T * r_{RP} = Fu * r_{RP}. \tag{14}$$

By replacing Equations (10), (11), (13) and (14) in Equation (12), it results:

$$Fu * r_{RP} = mg * u_1 + \left(m\ddot{x} + k_a\dot{x}^2\right) * r_{W1} + Ce. \tag{15}$$

By rewriting Equation (15), the acceleration can be evaluated as:

$$\ddot{x} = \frac{Fu}{\tau m} - \frac{k_a}{m r_{W1}}\dot{x}^2 - \frac{mg * u_1}{m r_{W1}} - \frac{Ce}{m r_{W1}}. \tag{16}$$

Equation (16) and the model of the user's power have been implemented in Matlab/Simulink.

During the steady-state phase, the average wheelchair speed is constant. Figure 16 show respectively the wheelchair speed and the user's power during a complete cycle: traction phase and recovery phase.

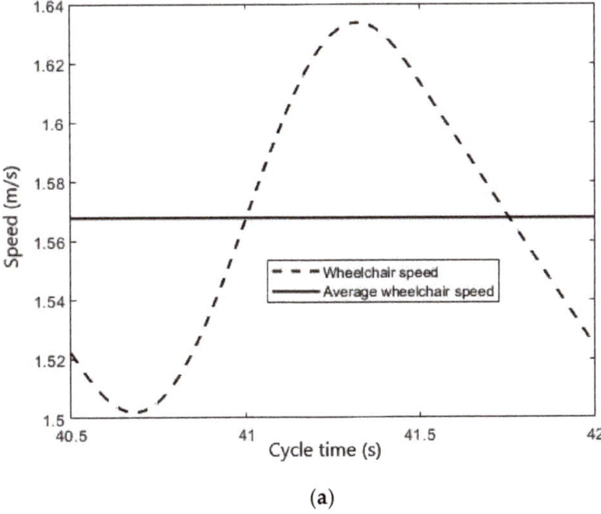

(a)

Figure 16. *Cont.*

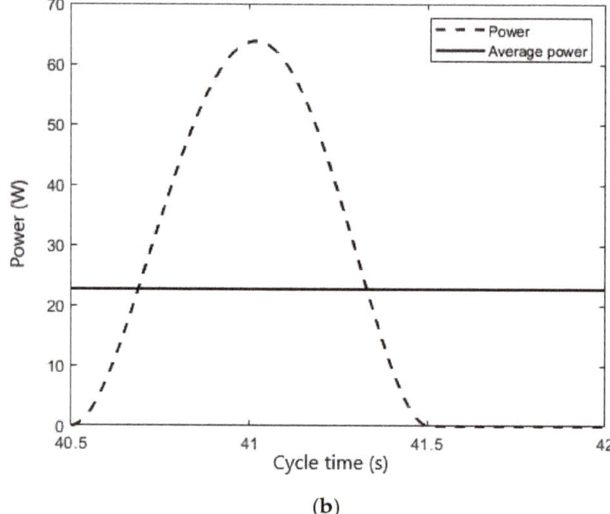

(b)

Figure 16. Wheelchair speed (**a**) and user's power (**b**) in a complete cycle during the steady-state phase.

The average power of 23 W and the average speed of 1.57 m/s = 5.65 km/h obtained with the simulation is in accordance with the tests carried out in [12,13,15].

By employing the user's force model, we obtain a Fu_max = 120 N as shown in Figure 17, while the average force Fu_avg = 43 N. The force Fu is the sum of the right and left arm contribution: $Fu_r + Fu_l$. Then, the average force for each arm is approximately

$$\overline{Fu_r} = \overline{Fu_l} = 22 \text{ N}. \tag{17}$$

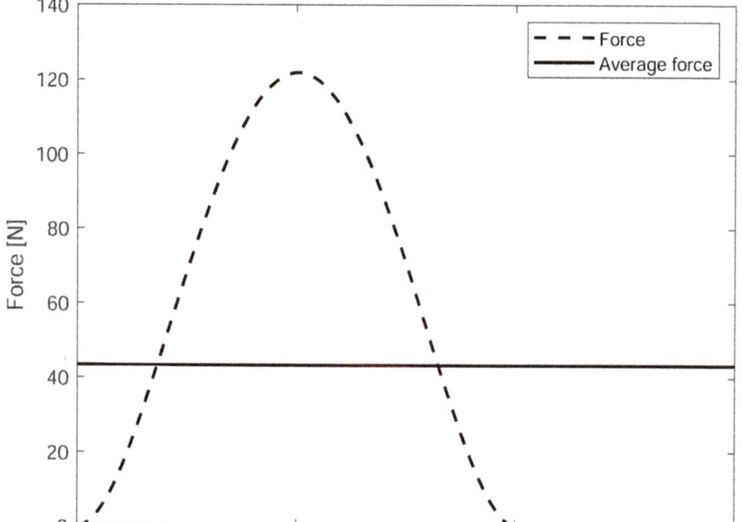

Figure 17. User's force during the steady-state phase.

6. Conclusions

The development of an innovative system of propulsion for manual wheelchairs is described in different possible sketches in order to be implemented in sports wheelchairs and for wheelchairs employed in everyday life. In addition, the design of the mechanism for the transmission of motion is presented. The innovative system of propulsion could be a robust idea to solve injuries on the upper limb caused by the other manual system of propulsion. The project of the prototype, named Handwheelchair.q02, is still being developed.

A simplified dynamic model has provided preliminary results in terms of force and power during a rowing stroke cycle applied to a manual wheelchair. In the future, the test data of the prototype Handwheelchair.q02 will confirm or deny these predictions.

Author Contributions: Conceptualization, G.Q. and E.B.; methodology, G.Q., E.B., P.C.; software, E.B. and P.C.; validation, G.Q.; investigation, P.C.; data curation, P.C.; writing—original draft preparation, P.C.; writing—review and editing, G.Q., E.B. and P.C.; visualization, P.C.; supervision, G.Q.; project administration, G.Q.

Funding: This research received no external funding.

Conflicts of Interest: The authors declare no conflict of interest.

References

1. Lewis, A.R.; Philips, E.J.; Robertson, W.S.P.; Grimshaw, P.N.; Portus, M. Injury prevention of elite wheelchair racing athletes using simulation approaches. *Proceedings* **2018**, *2*, 255. [CrossRef]
2. Curtis, K.A.; Roach, K.E.; Applegate, E.B.; Amar, T.; Benbow, C.S.; Genecco, T.D.; Gualano, J. Reliability and validity of the wheelchair user's shoulder pain Index (WUSPI). *Paraplegia* **1995**, *33*, 595–601. [CrossRef]
3. Cooper, R.A.; Boninger, M.L.; Robertson, R.N. Repetitive strain injury among manual wheelchair users. *Team Rehab. Rep.* **1998**, *9*, 35–38.
4. Van Der Woude, L.H.V.; Dallmeijer, A.J.; Jansenn, T.W.J. Alternative modes of manual wheelchair ambulation: An overview. *Am. J. Phys. Med. Rehabil.* **2001**, *80*, 765–777. [CrossRef]
5. Cooper, R.A.; De Luigi, A.J. Adaptive sports technology and biomechanics: Wheelchairs. *Paralympic Sports Med. Sci.* **2014**, *6*, 31–39. [CrossRef]
6. Rice, I. Recent salient literature pertaining to the use of technology in wheelchair sports. *Curr. Phys. Med. Rehabil. Rep.* **2016**, *4*, 329–335. [CrossRef]
7. Van Delen, C.J.R.T.; Gooch, S.; Ingram, B.J.; Borren, G.L.; Jenkins, A.; Dunn, J. Classification efficiency in wheelchair rugby: Strength analysis. *Int. Fed. Autom. Control* **2014**, *47*, 9901–9906. [CrossRef]
8. Gil-Agudo, A.; Del Ama-Espinosa, A.; Crespo-Ruiz, B. Wheelchair basketball quantification. *Phys. Med. Rehabil. Clin. N. Am.* **2010**, *21*, 141–156. [CrossRef]
9. Fung, Y.; Chan, D.K.; Caudwell, K.M.; Chow, B. Is the wheelchair fencing classification fair enough? A kinematic analysis among world-class wheelchair fencers. *Eur. J. Adapt. Phys. Act.* **2013**, *6*, 17–29. [CrossRef]
10. Guan, T.; Lei, L.; Li, J. The human engineering analysis of the racing wheelchair. *Adv. Mater. Res.* **2014**, *538–541*, 2802–2806. [CrossRef]
11. Litzenberger, S.; Mally, F.; Sabo, A. Biomechanics of elite recumbent handcycling: A case of study. *Sport Eng.* **2016**, *19*, 201–211. [CrossRef]
12. Arnet, U.; Van Drongelen, S.; Van der Woude, L.H.V.; Veeger, D.H.E.J. Shoulder load during handcycling at different incline and speed conditions. *Clin. Biomech.* **2012**, *27*, 1–6. [CrossRef]
13. Arnet, U.; Van Drongelen, S.; Scheel-Sailer, A.; Van der Woude, L.H.V.; Veeger, D.H.E.J. Shoulder load during synchronous handcycling and handrim wheelchair propulsion in persons with paraplegia. *J. Rehabil. Med.* **2012**, *44*, 222–228. [CrossRef]
14. Bregman, D.J.J.; Van Drongelen, S.; Veeger, H.E.J. Is effective force application in handrim wheelchair propulsion also efficient? *Clin. Biomech.* **2009**, *24*, 13–19. [CrossRef]
15. Sasaki, M.; Stefanov, D.; Ota, Y.; Miura, H.; Nakayama, A. Shoulder joint contact force during lever-propelled wheelchair propulsion. *Robomech J.* **2015**, *2*, 13. [CrossRef]

16. Greene, A.J.; Sinclair, P.J.; Dickson, M.H.; Colloud, F.; Smith, R.M. The effect of the ergometer design on a rowing stroke mechanics. *Scand. J. Med. Sci. Sports* **2013**, *23*, 468–477. [CrossRef]
17. Pelz, P.F.; Verge, A. Validated biomechanical model for efficiency and speed of rowing. *J. Biomech.* **2014**, *47*, 3415–3422. [CrossRef]
18. Smoljanovic, T.; Bojanic, I.; Pollock, C.L.; Radonic, R. Rib stress fracture in a male adaptive rower from the arms and shoulders sport class: Case report. *Croat. Med. J.* **2011**, *52*, 644–647. [CrossRef]
19. Quaglia, G.; Bonisoli, E.; Cavallone, P. A proposal of alternative propulsion system for manual wheelchair. *Int. J. Mech. Control* **2018**, *19*, 33–38.
20. Quaglia, G.; Nisi, M. Design of a self-leveling cam mechanism for a stair climbing wheelchair. *Mech. Mach. Theory* **2017**, *112*, 84–104. [CrossRef]
21. Quaglia, G.; Franco, W.; Nisi, M. Design of a reconfiguration mechanism for an electric stair-climbing wheelchair. In Proceedings of the ASME IMECE 2014 International Mechanical Engineering Congress and Exposition, Montreal, QC, Canada, 14–20 November 2014; p. V04AT04A022.
22. Baldissera, P. Proposal of a coast_down model including speed-dependent coefficients for the retarding forces. *Proc. Inst. Mech. Eng. Part P* **2017**, *231*, 154–163.
23. Turpin, N.A.; Guèvel, A.; Durand, A.; Hug, F. Effect of power output on muscle coordination during rowing. *J. Appl. Physiol.* **2011**, *111*, 3017–3029. [CrossRef]

Article

Development and Testing of a Low-Cost Wireless Monitoring System for an Intelligent Tire[†]

Giovanni Breglio [1], Andrea Irace [1], Lorenzo Pugliese [2], Michele Riccio [1], Michele Russo [1], Salvatore Strano [2,*] and Mario Terzo [2]

[1] Department of Electrical Engineering and Information Technologies, University of Naples Federico II, Via Claudio 21, 80125 Naples, Italy
[2] Department of Industrial Engineering, University of Naples Federico II, Via Claudio 21, 80125 Naples, Italy
* Correspondence: salvatore.strano@unina.it
† This paper is an extended version of our paper published in Breglio, G.; Irace, A.; Pugliese, L.; Riccio, M.; Russo, M.; Strano, S. and Terzo, M. Cost-Effective Wireless Sensing System for an Intelligent Pneumatic Tire. In the proceedings of the International Conference of IFToMM ITALY, Cassino, Italy, 29–30 November 2018.

Received: 5 April 2019; Accepted: 26 June 2019; Published: 8 July 2019

Abstract: Intelligent tire concept constitutes one of the approaches to increase the accuracy of active safety systems in vehicle technology. The possibility of detecting tire–road interactions instantaneously has made these systems one of the most important research areas in automotive engineering. This study introduces the use of cost-effective flex and polyvinylidene fluoride strain sensors to estimate some dynamic tire features in free-rolling and real working conditions. The proposed solution combines a microcontroller-based readout circuit for the two sensors with a wireless data transmission system. A suitable prototype was realized and first experimental tests were conducted, in the laboratory as well as on the road. The energy consumption of the wireless monitoring system was optimized. Simulated and experimental results validate the proposed solution.

Keywords: Intelligent tire; flex sensor; PVDF sensor; energy saving; vehicle dynamics; smart systems

1. Introduction

Electronic controls are necessary for road safety these days: radars, electronic stability controls (ESC), active breaking systems (ABS) and so on. All the currently-used controls are based on on-board sensors. This means that most of the important parameters for the vehicle dynamics are estimated indirectly. Because of this disconnection between sensors and features of interest, imperfections and delays are generated in the metering system and so an incorrect or late vehicle control response is inevitable. Many researchers have realized that to improve the electronic controls, measurements should be made directly on the tire, as tires are the only vehicle part in contact with the road surface and all the road disturbances and forces act on them.

They called this new technology "Smart Tire" or "Intelligent Tire" to emphasize the tire autonomy in measuring its own forces, deformations, and accelerations and then communicate them wirelessly [1]. This aspect could be also crucial for the design of vehicle diagnostic systems based on modern signal processing techniques [2,3] and self-learning methods [4].

The research in Smart Tire applications began in the early 1980s with the "tire pressure monitoring system" (TPMS), which allows the driver to know the tire pressure conditions. Many steps have been made in this area and many different ideas have followed since. There are various adopted sensors, from optical sensors in [5], segmented capacitance rings for measuring tire strain in [6], ultrasonic sensors in [7] for contact patch deformations, polyvinylidene fluoride (PVDF) sensors to measure the strain [8,9], and electrical capacitance change of tire linked to the strain [10,11]. In [12], spectral features of the capacitance output signal are used to improve the accuracy of the strain measurement.

In addition, a new type of capacitance sensor based on ultra-flexing epoxy resin is shown in [13] and a Hall effect based magnetic sensor is used in [14] to measure the thread deformation. In [15–17], algorithms for the estimation of slip angle and tire working conditions from strain measurements were presented.

This paper aims to validate a new approach to intelligent tire solutions using flex and PVDF sensors with a low-power wireless communication system and a microcontroller readout unit. Differently from common approaches, based on acceleration measurements, this activity employees strain measurements to make the tire intelligent. This choice is due to the high sensitivity to noise of the accelerometers. Moreover, this paper presents a new approach characterized by two different strain measurements provided by two different sensor technologies: the flex sensor suitable for static conditions and low dynamics and the PVDF suitable for high dynamics.

2. Intelligent Tire System

The prototype, based on a flex sensor and a PVDF sensor, was used to monitor a 245/40/R18 tire. The flex sensor (see Figure 1a,b), due to a series of carbon resistive elements, works like a variable resistance from a constant value for a flat sensor to a maximum value in a totally bended condition (90°). Once attached to the inner part of the tire, the sensor can be used to estimate the curvature variation of the tire due to the bending action, which can be then linked to the working conditions (e.g., rolling speed, vertical load, and length of the contact patch).

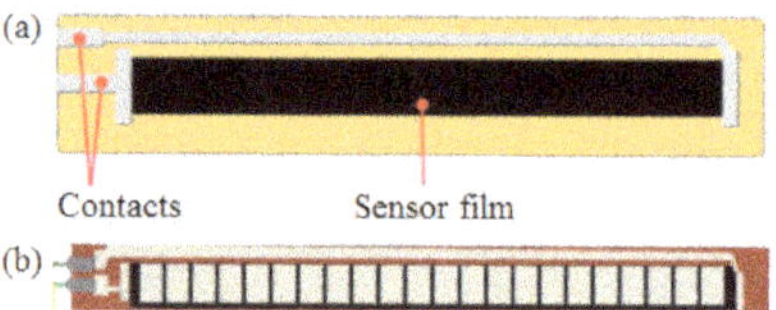

Figure 1. (**a**) Scheme of a resistive flex sensor; top view: electrical contacts in grey, conductive film in black; (**b**) Flex sensor by Inage SI Inc. Courtesy of Flexpoint Inc.

Meanwhile the PVDF sensor (Figure 2a) is based on a piezo-electric material, therefore it can generate an electric charge proportional to the applied mechanical deformations.

A schematic picture of the proposed electronic system is depicted in Figure 2a. The data acquisition system (DAS) implemented is mainly composed of two conditioning circuits, a microcontroller and a wireless communication module.

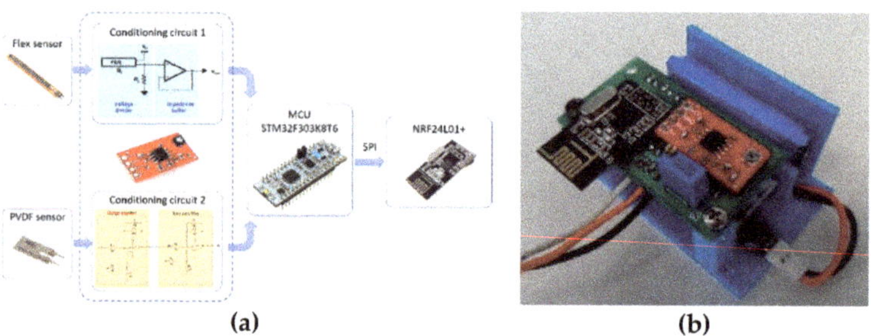

Figure 2. (**a**) Block diagram of the DAS (data acquisition system) for smart tire sensors; (**b**) The prototype equipped with a 500 mAh/3.3 V battery.

The microcontroller unit (MCU) has two fast 12-bit ADCs with 5 Msps maximum sampling rate, while in our application the sampling frequency is $f_S = 12$ kHz. In addition, the developed DAS can be configured to retrieve data from the sensor both in real-time and off-line. In the first case, the microcontroller sends data packets over a Bluetooth channel to a personal computer; in the second case, the microcontroller collects the signal data so that all the information and numerical data can be stored in a non-volatile SD card through serial interface.

The MCU firmware is written with the Arm Mbed IoT Device Platform and it is based on the Mbed OS 5.8 release, whereas the synchronization between the sampling operation and the writing operation is ensured by interrupt/tread mechanisms. Finally, the wireless communication has been provided adopting the nRF24L01 module.

In Figure 3, it is possible to see how the sensors and the sensing circuit, held by a 3D printed base, have been mounted on the inner part of the tire. An elastic support fixes the sensing circuit base to the tire inner liner.

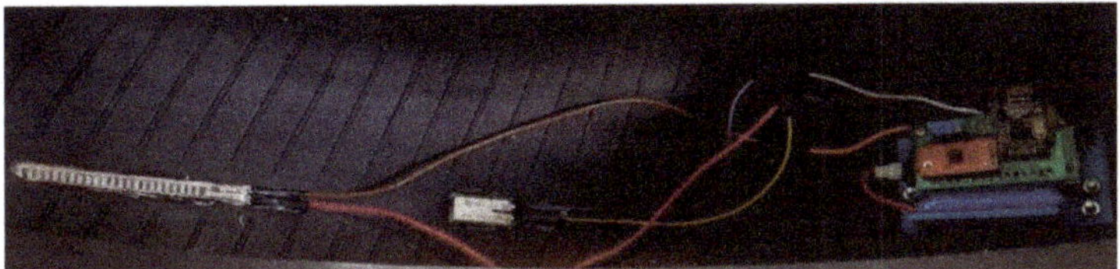

Figure 3. Sensing circuit, flex and PVDF sensors mounted into the tire.

2.1. Energy Consumption Optimisation

As the sensing circuit will work in autonomy inside the tire, a lot of effort has been put into reducing its energy consumption and trying to create a system as long lasting as possible.

The most inefficient part of the system is the wireless communication, so starting from the firmware, a "power down" mode has been introduced to interrupt the communications whenever the sensor gives no signal. Once the signal is different from the reference value, the microcontroller starts the ISR (interrupt service routine), waking-up the system and allowing communication. Thanks to these modifications, a consumption analysis revealed a decrease in electricity consumption of about 33.4% (from 564 to 169 mAh), while the stand-by consumption has been lowered to 0.46 µAh.

As one would expect, the battery has an important role in the energetic autonomy of the system, at the moment the sensing circuit is powered by a 3.7 V and 500 mAh LiPo battery. In the near future, it will be replaced by a "button cell", which will also allow for a considerable miniaturization of the system.

2.2. Characteristics of the Flex and PVDF Sensor Signals

To properly understand how the flex sensor is affected by the tire motion, we should follow its passage through the contact patch. Firstly, the sensor enters the contact patch and starts to be deformed simultaneously with the tire tread band (See Figure 4a) until the moment in which it is placed exactly in the contact patch. When that happens its curvature decreases, reaching a minimum. Finally, during the exit from the contact patch the sensor curvature increases again, due to the tire tread band deformation. The radius of curvature and its time variation behaviour, in a free-rolling condition, is shown in Figure 4c.

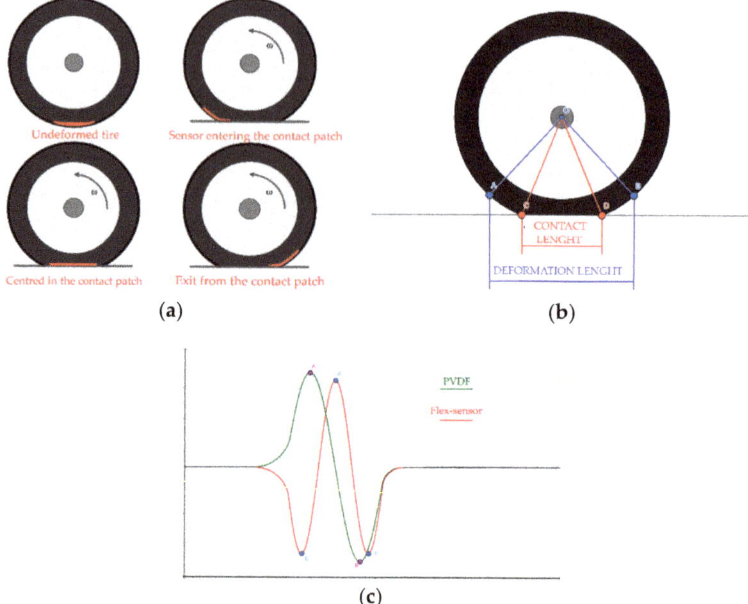

Figure 4. (**a**) Flex sensor deformation during tire rolling; (**b**) difference between contact length and deformation length; (**c**) waveforms of the radius of curvature and its time variation.

In the absence of braking, traction or steering forces, the radius of curvature is almost symmetric with respect to the contact patch center (the point D in Figure 4c). The time variation of the radius of curvature related to the flex sensor can be adopted for estimating the length of the contact patch as shown in Figure 4b. On this last consideration, it is necessary to point out the difference between the contact length and deformation length (Figure 4b). As the PVDF returns the derivative form of the flex-sensor output, due to this link between the two sensors output, considerations about the flex-sensor could be applied for the PVDF as well (considering that a variation in value for the flex-sensor output matches with a slope variation for the PVDF). As the preliminary results have been obtained at low tire rolling speed, the PVDF outputs have been not analyzed in detail but only as a verification of the physical meaning of the flex sensor signals. Therefore, the mathematical modelling and the main experimental results are referred to the flex sensor.

2.3. Mathematical Modelling of the Flex Sensor Signal

In this section, a mathematical description of the flex sensor measurement is described. The modelling approach is based on the "flexible ring tire model" [18] (Figure 5).

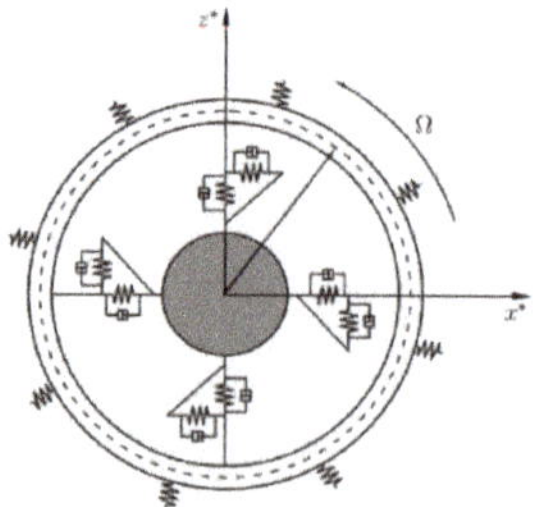

Figure 5. Flexible ring tire model.

As shown in Figure 5, the treadband structure is modelled as a thin circular elastic ring connected to the wheel hub in circumferential and radial directions through a viscoelastic foundation. Assuming the tire tread band is inextensible, the tire motion equations can be written using the relationships between the radial and tangential ring displacements (−w and v) in Equation (1), the circumferential strain $\epsilon_{\theta\theta}$ and the tire inner liner (Equation (2)):

$$- w = \frac{\partial v}{\partial \theta} \tag{1}$$

$$\epsilon_{\theta\theta} = \frac{h}{2} k(\theta) = -\frac{h}{2R^2}\left(\frac{\partial v}{\partial \theta} - \frac{\partial^2 w}{\partial \theta^2}\right) \tag{2}$$

where θ is the angular coordinate of a point belonging to the treadband in the fixed-body reference frame, h is the ring thickness, R is the undeformed ring mean radius and $k(\theta)$ is the local curvature. Thanks to the algorithm illustrated in [19], it is possible to evaluate the tangential and radial displacements in closed form and consequently the local tire curvature.

The flex-sensor circuit model is based on the sensor itself, a variable resistance which follows Equation (3) and the voltage divider characterized by Equation (4) (see Figure 2 for reference).

$$R_2(t) = R_0 + ck(t) \tag{3}$$

$$V_{OUT} = V_{IN}\frac{R_2(t)}{R_1 + R_2(t)} \tag{4}$$

An example of behavior of the voltage from the flex sensor is presented in Figure 6.

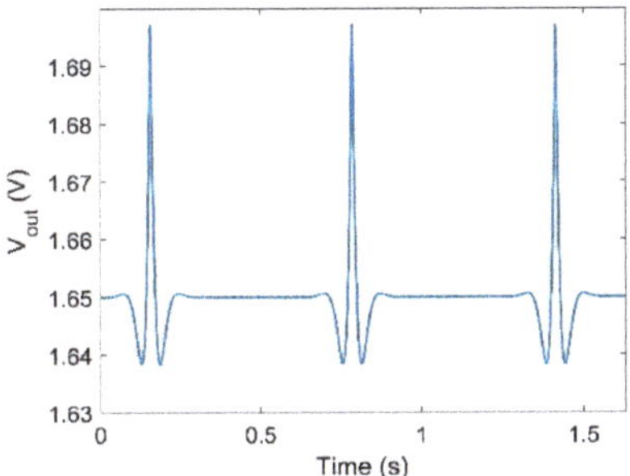

Figure 6. Simulated flex-sensor voltage output.

2.4. Tire Test Rig

The test rig used during the tire testing session (shown in Figure 7) is composed of a portal frame that works as support of the wheel axle.

The wheel axle is mounted on two vertical linear guides in order to transmit the vertical load to the pneumatic tire. On the top of the portal frame, there is a housing for the loading cell, which allows measurement of the vertical applied load. A hydraulic cylinder is then placed between the load cell and a horizontal beam, which is directly connected to the guides. The measurements provided by the flex sensor were obtained for different values of vertical loads and tire rolling speeds. This configuration allowed us to check if the output signals from the sensors were in accordance with their theoretical descriptions.

Figure 7. Tire test rig.

3. Experimental Results

The first testing session focused on the flex sensor signal behavior at low speed (up to 20 Kph); the same test was repeated three times (in Figure 8a), with each one in different load conditions to show how the sensor signal is affected. The peak value increases with the applied load in the order of 70 mV/50 kg, this result allows the system to estimate the applied load. Figure 8b shows the variation of the maximum amplitude value with respect to the vertical load (Figure 8a).

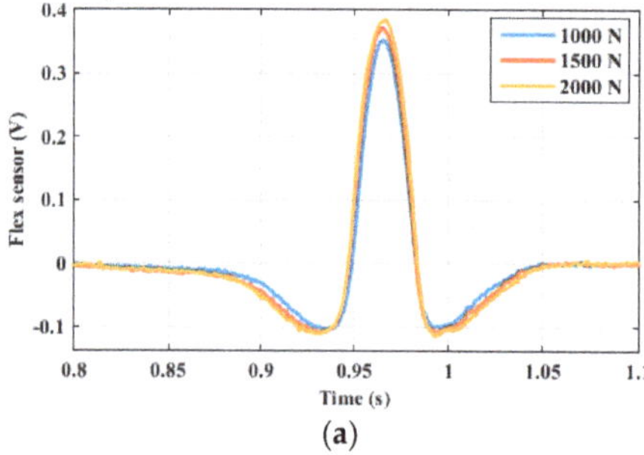

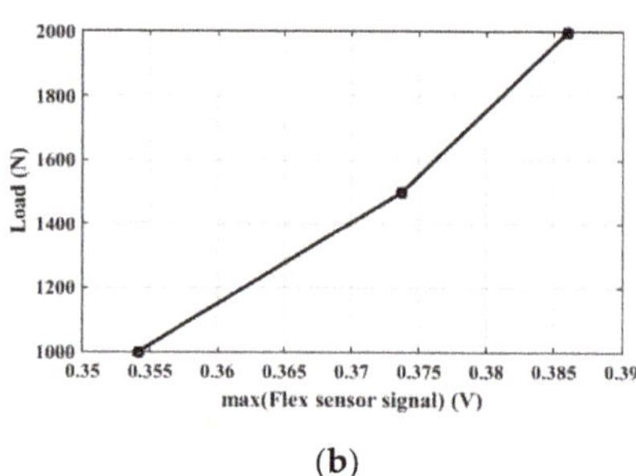

Figure 8. (**a**) Measurements for different vertical loads (tire rolling speed: 10 km/h); (**b**) vertical load for different maximum values of the sensor signal.

The signals (in Figure 8a) have a qualitative behavior similar to the theoretical radius of curvature signal presented in Figure 4c, this is an important feedback about the system trustworthiness.

The Figure 9a shows a different testing session, based on the same principle. It is obtained by maintaining a constant tire vertical load (2000 N) and varying the wheel rolling velocity in order to identify and check only the speed effects. Figure 9b relates the wheel rolling velocity for different values of the sensor rotational speed, obtained from the time period between two consecutive positive peaks of the sensor output.

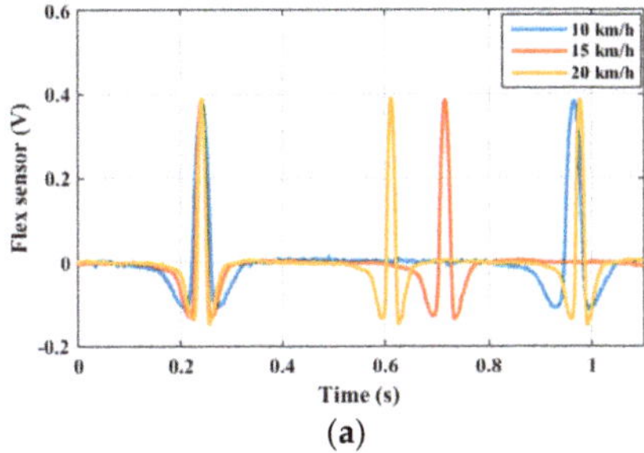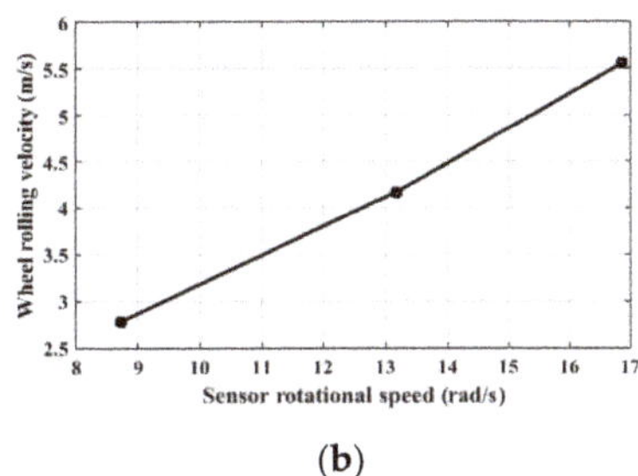

(a) **(b)**

Figure 9. (**a**) Measurements for different wheel rolling velocities (vertical load 2000 N); (**b**) wheel rolling velocity for different values of the sensor rotational speed.

The experimental results presented in Figure 9a,b demonstrate the possibility to obtain an estimation of the wheel rolling velocity from a signal processing method applied to the wireless measurements. Furthermore, under the hypothesis of a perfect bonding between flex sensor and tire inner liner, the sensor rotational speed coincides with the tire carcass speed. The results of Figure 9b show a linear relationship between the sensor rotational speed and the wheel rolling velocity.

The PVDF also provides a signal (shown in Figure 10) qualitatively similar to the expected output (Figure 4c).

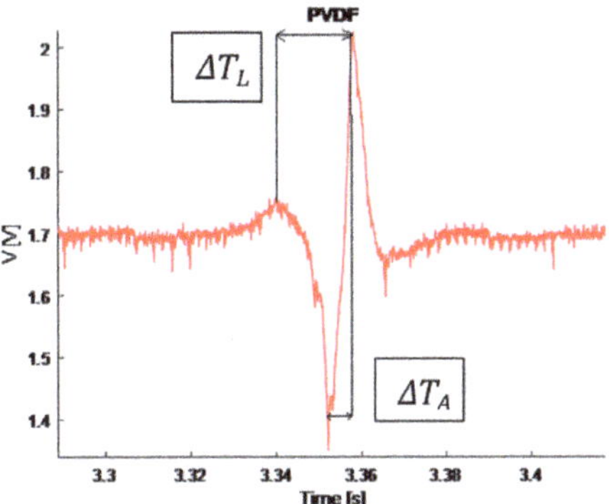

Figure 10. PVDF experimental output signal.

Once these tests were completed the research focused on the effects of steering on the signals, mainly from the flex-sensor, in a series of road tests. For instance, Figure 11 shows a series of events where the intelligent tire "feels" a vertical load variation due to a steering correction.

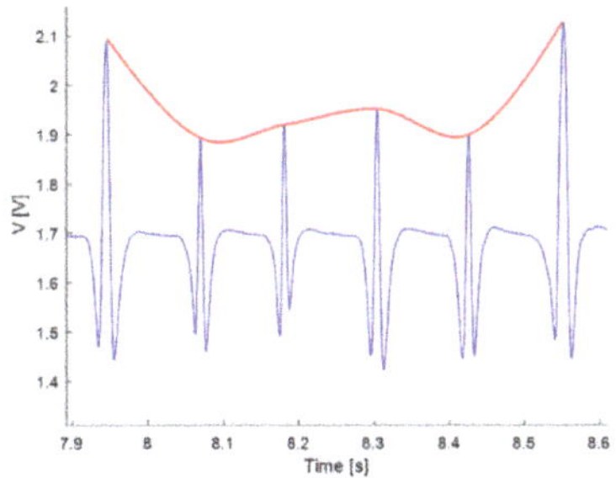

Figure 11. Road tests: flex sensor signal for a vertical load variation.

Another important achievement was reached in the road test session: the possibility to estimate the vehicle acceleration and braking actions. Both of them affect the two lower peaks of the flex sensor signal according to the action strength (Figure 12).

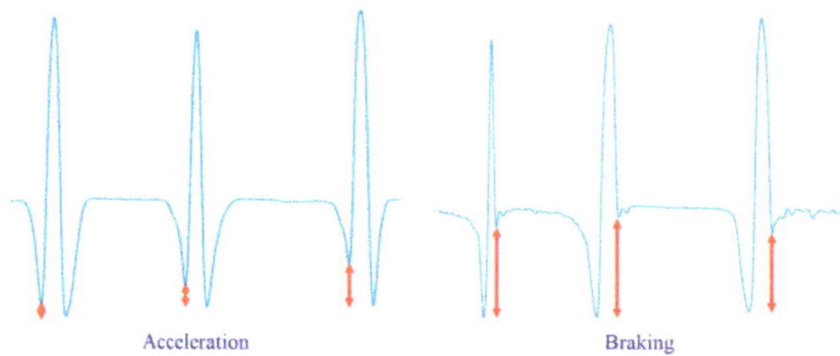

Figure 12. Acceleration and braking effects on the flex-sensor output signal.

4. Discussion

The test sessions successfully returned the expected theoretical results obtained by studying the "flexible ring tire model". Starting from the laboratory, the tests focused on the relationship between the output from the sensors and features of interest. Most of the job focused on the flex-sensor, and as a consequence, it has been possible to anticipate what could be discovered in road tests, whereas the PVDF-sensor played the role of controlling the truthfulness of the flex-sensor readings. So, on one hand the prototype has been tested and controlled on its use, on the other hand it has gradually been miniaturized in order to not interfere significantly with the tire motion. The system energy requirement was lowered, by working on the firmware, by 33.4%. Furthermore, improvements were obtained by finding the best compromise between economy, energy consumption and performance of the wireless communication system.

5. Conclusions

In this paper, a cost-effective wireless system for an intelligent tire prototype is presented. The proposed apparatus mainly consists in a sensing circuit equipped with a micro-controller unit, low-power wireless communication module, a low-cost flex sensor and a PVDF sensor to detect the strain rate. A tire model was analysed to predict the flex sensor output.

Subsequently, an experimental activity was conducted in order to analyze the feasibility of the proposed approach. In particular, the test rig was adopted to evaluate the possibility of estimating some tire working condition features from the flex sensor and the PVDF in free-rolling conditions and low speeds.

The research continued with a series of road tests, in which the intelligent tire prototype was tested in cornering, acceleration and breaking in real conditions. Strategies to reduce the sensing circuit energy consumption were investigated. Correlations between measurements and physical parameters were investigated as preliminary analysis. Further improvements of this estimation procedure will include an extensive experimental activity oriented to validate the proposed method with more experimental data.

Author Contributions: Conceptualization, S.S., M.T.; methodology, G.B., A.I., M.Ri.; software, M.Ri.; data curation, M.Ri., L.P., S.S.; writing—original draft preparation, L.P.; writing—review and editing, G.B., A.I., M.Ri., S.S., M.T.; supervision, M.Ru.

Funding: This research received no external funding.

Acknowledgments: The authors would like to thank Giuseppe Iovino and Gennaro Stingo for their technical support.

Conflicts of Interest: The authors declare no conflict of interest.

References

1. Lee, H.; Taheri, S. Intelligent tires-a review of tire characterization literature. *IEEE Intell. Transp. Syst. Mag.* **2017**, *9*, 114–135. [CrossRef]
2. Acosta, M.; Kanarachos, S.; Blundell, M. Virtual tyre force sensors: An overview of tyre model-based and tyre model-less state estimation techniques. *Proc. Inst. Mech. Eng. Part D J. Automob. Eng.* **2018**, *232*, 1883–1930. [CrossRef]
3. Acosta, M.; Kanarachos, S.; Blundell, M. Road friction virtual sensing: A review of estimation techniques with emphasis on low excitation approaches. *Appl. Sci.* **2017**, *7*, 1230. [CrossRef]
4. Niola, V.; Quaremba, G.; Savino, S. The objects location from images binarized by means of self-learning neural network. *WSEAS Trans. Syst.* **2005**, *4*, 417–423.
5. Tuononen, A. Optical position detection to measure tyre carcass deflections. *Veh. Syst. Dyn.* **2008**, *46*, 471–481. [CrossRef]
6. Cullen, J.; Arvanitis, N.; Lucas, J.; Al-Shamma'a, A. In-field trials of a tyre pressure monitoring system based on segmented capacitance rings. *Measurement* **2002**, *32*, 181–192. [CrossRef]
7. Magori, V.; Magori, V.R.; Seitz, N. On-Line Determination of Tyre Deformation, a Novel Sensor Principle. In Proceedings of the IEEE Ultrasonics Symposium, Sendai, Japan, 5–8 October 1998.
8. Moona, K.S.; Liangb, H.; Yi, J.; Mika, B. Tire tread deformation sensor and energy harvester development for "Smart Tire" applications. In Proceedings of the Sensors and Smart Structures Technologies for Civil, Mechanical, and Aerospace System, San Diego, CA, USA, 19–22 March 2007.
9. Yi, J. A piezo-sensor-based 'smart tire' system for mobile robots and vehicles. *IEEE/ASME Trans. Mechatron.* **2008**, *13*, 95–103. [CrossRef]
10. Todoroki, A.; Miyatani, S.; Shimamura, Y. Wireless strain monitoring using electrical capacitance change of tire: Part I—With oscillating circuit. *Smart Mater. Struct.* **2003**, *12*, 403–409. [CrossRef]
11. Todoroki, A.; Miyatani, S.; Shimamura, Y. Wireless strain monitoring using electrical capacitance change of tire: Part II—Passive. *Smart Mater. Struct.* **2003**, *12*, 410–416. [CrossRef]
12. Matsuzaki, R.; Todoroki, A. Passive wireless strain monitoring of actual tire using capacitance–resistance change and multiple spectral features. *Sens. Actuators A* **2006**, *126*, 277–286. [CrossRef]
13. Matsuzaki, R.; Todoroki, A. Wireless flexible capacitive sensor based on ultra-flexible epoxy resin for strain measurement of automobile tires. *Sens. Actuators A* **2007**, *140*, 32–42. [CrossRef]
14. Yilmazoglu, O.; Brandt, M.; Sigmund, J.; Genc, E.; Hartnagel, H.L. Integrated InAs/GaSb 3D magnetic field sensors for 'the intelligent tire. *Sens. Actuators A* **2001**, *94*, 59–63. [CrossRef]
15. Garcia-Pozuelo, D.; Olatunbosun, O.; Yunta, J.; Yang, X.; Diaz, V. A novel strain-based method to estimate tire conditions using fuzzy logic for intelligent tires. *Sensors* **2017**, *17*, 350. [CrossRef] [PubMed]
16. Garcia-Pozuelo, D.; Yunta, J.; Olatunbosun, O.; Yang, X.; Diaz, V. A strain-based method to estimate slip angle and tire working conditions for intelligent tires using fuzzy logic. *Sensors* **2017**, *17*, 874. [CrossRef] [PubMed]
17. Yang, X.; Olatunbosun, O.; Garcia-Pozuelo Ramos, D.; Bolarinwa, E. *FE-Based Tire Loading Estimation for Developing Strain-Based Intelligent Tire System*; SAE: Detroit, MI, USA, 2015.
18. Gong, S. *A Study of In-Plane Dynamics of Tires*; Delft University: Delft, The Netherlands, 1993.
19. Garcia-Pozuelo, D.; Olatunbosun, O.; Strano, S.; Terzo, M. A real-time physical model for strain-based intelligent tires. *Sens. Actuators A Phys.* **2019**, *288*, 1–9. [CrossRef]

Article

Investigation on the Mechanical Properties of MRE Compounds [†]

Renato Brancati *, Giandomenico Di Massa and Stefano Pagano

Dipartimento di Ingegneria Industriale, Università di Napoli Federico II, Via Claudio 21, 80125 Napoli, Italy; gdimassa@unina.it (G.D.M.); stefano.pagano@unina.it (S.P.)

* Correspondence: renato.brancati@unina.it; Tel.: +39-81-7683683

† This paper is an extended version of our conference paper Brancati, R.; Di Massa, G.; Di Vaio, M.; Pagano, S.; Santini, S. Experimental Investigation on Magneto-Rheological Elastomers. In Proceedings of the International Conference of IFToMM ITALY, Cassino, Italy, 29–30 November 2018.

Received: 6 March 2019; Accepted: 24 May 2019; Published: 1 June 2019

Abstract: This paper describes an experimental investigation conducted on magneto-rheological elastomers (MREs) with the aim of adopting these materials to make mounts to be used as vibration isolators. These materials, consisting of an elastomeric matrix containing ferromagnetic particles, are considered to be smart materials, as it is possible to control their mechanical properties by means of an applied magnetic field. In the first part of the paper, the criteria adopted to define the characteristics of the material and the experimental procedures for making samples are described. The samples are subjected to a compressive static test and are then, adopting a testing machine specially configured, tested for shear periodic loads, each characterized by a different constant compressive preload. The testing machine is equipped with a coil, with which it is possible to vary the intensity of the magnetic field crossing the sample during testing to evaluate the magneto-rheological effect on the materials' characteristics in terms of stiffness and damping.

Keywords: magneto-rheological elastomers; smart materials; semi-active isolator; iron powder

---

1. Introduction

Magneto-rheological elastomers (MREs) are capable of changing their stiffness and damping in response to an applied magnetic field [1,2]. This characteristic allows for the development of controllable devices for vibration isolation [3,4], overcoming the limits of traditional passive isolators. In fact, passive isolators shift the system's natural frequency far enough away from the range of the forcing frequencies to avoid resonance phenomena [5]. These types of isolators perform their tasks efficiently in machines operating in steady-state conditions. However, if the forcing frequencies are not known a priori or if the resonance condition is frequently crossed, the ability to change the isolators' characteristics in real time can lead to a significant improvement isolation.

MRE compounds are composed of an elastomeric matrix containing magnetizable particles of nano to micro sizes. During the curing phase, the compound is subjected to the action of a magnetic field to rearrange and to orient the ferromagnetic particles along the strength lines of the magnetic field. The materials' characteristics, in particular the stiffness, vary with the intensity of the magnetic field, depending on the particles' dimensions [6,7] and their rearrangement in the matrix.

The adoption of MRE isolators [8,9] may be particularly useful for solving machinery vibration isolation problems, as MRE isolators combine the reliability of passive devices with the ability of active devices by adapting their characteristics to the actual machine and environment conditions.

In this paper, an analytical model is presented that describes the influence of the volume fraction of magnetic particles on the magneto-rheological effect, defined as the material shear modulus increment caused by a magnetic field, on MREs.

MRE samples were, therefore, prepared using silicone and polyurethane rubbers. To create an evident magneto-rheological effect, it is necessary to use an elastomeric matrix that is not particularly stiff; for this reason, our experiments were carried out on silicone samples.

These samples were prepared by adopting a nylon mold that allowed us to place permanent magnets at its extremities so that, during the polymerization phase, the iron-carbonyl particles mixed with the liquid silicone, aligning along the strength lines of the magnetic field and forming anisotropic chain-like structures that became locked in place upon the final cure.

To adopt MREs as vibration isolators, for light structures excited along the horizontal direction, the formed samples were tested with static compression loads and subjected to variable shear excitations. To carry out these dynamic tests, an experimental set-up was developed, in which an electrodynamic shaker was adopted to induce a shear load with assigned amplitude and frequency on two samples at a time. The samples were placed on the core of a coil so that it was possible to generate a magnetic field crossing the sample material. Thus, in this way, it was possible to investigate the effect of the load frequency at different levels of magnetic field intensity.

2. The Magneto-Rheological Effect

MREs are compounds containing magnetizable particles (with varying sizes of 3–10 μm) in a non-magnetic matrix. Their mechanical and rheological properties can be reversibly changed upon exposure to a suitable magnetic field [10]. The physical material property, which undergoes a significant variation depending on the strength of the magnetic field, is the shear elastic modulus; this characteristic allows for the realization of MRE pads with controllable stiffness. The magneto-rheological effect reflects the shear modulus change with respect to the value that it assumes in the absence of a magnetic field. MRE compounds may be isotropic if the particles are randomly dispersed or anisotropic if the particles are arranged in columns. Experimental tests have shown that the magneto-rheological effect is significantly larger in anisotropic MREs [11]. This arrangement can be obtained by immersing the liquid mixture in a static magnetic field during the vulcanization phase.

Figure 1 shows a schematic of a column of equally sized spherical particles, arranged at the same distance from each other, subjected to magnetic field H.

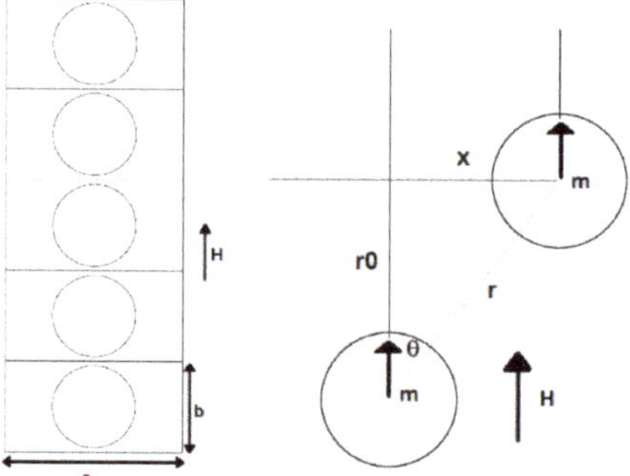

Figure 1. Particle column of an anisotropic magneto-rheological elastomer (MRE).

To estimate the magneto-rheological (MR) effect, a static model was developed. First, it must be noted that the compound shear modulus G, in the absence of a magnetic field, depends on the matrix module, G_0, and on the amount of iron powder. The following formula can be used to evaluate G as a function of the iron particles' volume fraction φ [12]:

$$G = G_0\left(1 + 1.25\phi + 14.1\phi^2\right) \tag{1}$$

To estimate the change in shear stress G_m due to the magnetic field, Jolly et al. proposed a dipole model [13]. This model is based on the hypothesis that all the particles are perfectly spherical and arranged in equidistant columns. Applying an external magnetic field parallel to the column, the inter-particle magnetic force induces added shear stress. The interaction energy of the two dipoles was calculated by Rosensweig [14] as follows:

$$E_{12} = \frac{1}{4\pi\mu_1\mu_0}\left[\frac{m_1 m_2 - 3(m_1 e_r)(m_2 e_r)}{|r|^3}\right] \tag{2}$$

where m_1 and m_2 are the magnetic dipoles strength, μ_0 is the vacuum permeability, and μ_1 is the relative permeability of the medium. The stress induced by the magnetic field can be calculated from the derivative of average energy density $U = n\,E_{12}/V$, where n is the particles' number and V is the volume occupied by the particles:

$$\sigma = \frac{\partial U}{\partial \epsilon} = \frac{9\phi\epsilon\left(4 - \epsilon^2\right)m^2}{2\pi^2\mu_1\mu_0 d^3 r^3 (1 + \epsilon^2)^{7/2}} \tag{3}$$

where $\epsilon = x/r_0$ is the shear strain of the particle column and d is the particle diameter. The magneto-induced shear storage modulus is as follows:

$$G_m = \frac{\partial\sigma}{\partial\epsilon} = \frac{\left(4\epsilon^4 - 27\epsilon^2 + 4\right)\phi J_p^2}{8\mu_1\mu_0 h^3 (1 + \epsilon^2)^{9/2}} \tag{4}$$

$$J_P = \mu_0 M_P \tag{5}$$

where J_P is the dipole moment magnitude per unit particle volume. The parameter $h = r_0/d$ gives an indication of the distance between the particles. Based on the schematic of Figure 1, $b = d{\cdot}h$, and the gap between the particles is equal to $d{\cdot}(h - 1)$. The particles are equally spaced if $a = b$ and the iron particles' volume fraction ϕ is

$$\phi = \frac{\frac{4}{3}\pi\frac{d^3}{8}}{d^3 h^3}| \rightarrow |h^3 = \frac{\pi}{6\phi}. \tag{6}$$

Then, remembering the law of Frohlich–Kennelly [15],

$$M_P = \frac{(\mu_P - 1)M_S H}{M_S + (\mu_P - 1)H} \tag{7}$$

where M_S is the saturated magnetization, H is the magnetic field strength, and μ_p is the relative permeability of the particles.

The following is therefore true:

$$G_m = \frac{6\left(4\epsilon^4 - 27\epsilon^2 + 4\right)\phi^2\mu_0}{8\mu_1\pi(1 + \epsilon^2)^{9/2}}\left(\frac{(\mu_P - 1)M_S H}{M_S + (\mu_P - 1)H}\right)^2 \tag{8}$$

Using Equations (1) and (8), it is possible to evaluate the MR effect, which is defined as follows:

$$\Delta G = \frac{G_m + G}{G}. \tag{9}$$

Figure 2 shows the MR effect as a function of the H field for $\phi = 0.27$ and for some values of the matrix shear modulus, G_0. The curves were obtained for iron saturation magnetization $M_S = 2.1$ T, $\mu_P = 1000$, and $\mu_1 = 1$ (elastomeric matrix). It can be observed that the MR effect is greater for lower values of G_0. The maximum values of ΔG, as a function of the powder volume fraction ϕ, are shown in Figure 3.

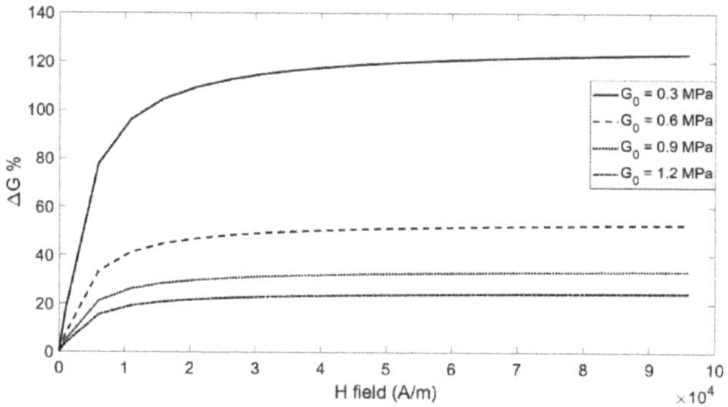

Figure 2. Magneto-rheological (MR) effect for some values of matrix moduli G_0.

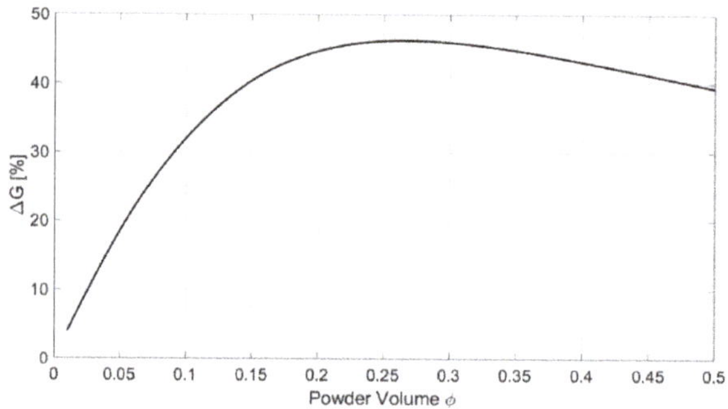

Figure 3. MR effect vs. powder volume fraction ϕ.

The MR effect reaches the maximum value for ϕ in the range of 0.25–0.30; with a greater quantity of powder, the MR effect decreases.

It is well known that the dynamic behavior of an elastomer differs from its static behavior due to its visco-elastic properties, which are characterized by hysteretic behavior. In order to study both static and dynamic behavior, magneto-rheological rubber samples were prepared for use in our experimental investigations.

3. MRE Sample Preparation

MRE matrixes are usually made of natural rubber or silicone rubber, as they have a low stiffness; therefore, the magnetorheological effect is more evident [16,17]. Here, the MRE samples were composed of a silicon elastomer matrix (Prochima GLS-10), characterized by 10 shore-A hardness, with micro iron-carbonyl particles (4–6 μm); the particles' volume percentage was equal to 25%.

The samples were formed in a nylon mold (Figure 4) (RS), in which the MRE liquid mixture was cured; two plastic diaphragms (DP) were adopted to separate the mixture from two neodymium permanent magnets, contained in plastic rings (RM) and characterized by a maximum energy product of 263–287 kJ/m^3. The magnetic field generated by the permanent magnets had the effect of orienting and aligning the particles according to the force lines of the magnetic field.

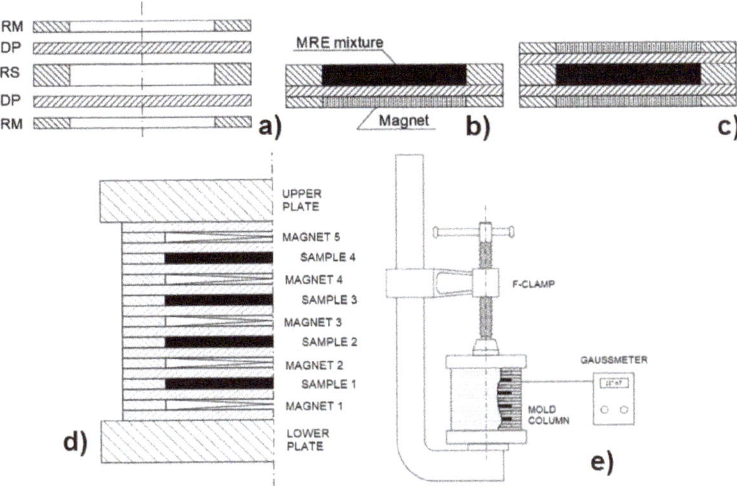

Figure 4. (**a**) Mold elements; (**b**) pouring phase; (**c**) polymerization phase; (**d**) multilayer mold; (**e**) static magnetic field measure.

The magnetic field intensity, measured by placing the probe of a gaussmeter (Brockhous BMG101) between the sample and the nylon diaphragm, was equal to 190 mT.

The mold containing the mixture was previously placed in a vacuum chamber, at about 50 mbar, twice for about 30 min to remove air bubbles from the mixture.

The vacuum chamber was composed of a steel pipe, which was closed at the ends with two plates; in the contact zone between the pipe and plates, there was an O-ring to ensure the airtightness [18]. When the vacuum pump was initiated, the plates were squeezed on the pipe, so no devices were required to connect the plates and pipe.

To prevent bonding between the MRE sample and the internal mold surface and to easily extract the formed sample from the mold, a wax-based release agent was used. The liquid mixture, which was poured into the mold, was cured at constant ambient temperature for about 24 h.

Figure 5 shows several depictions of the sample preparation phases. The sample diameter and height were equal to 50 and 6 mm, respectively. The main sample properties are reported in Table 1; it was noted that the shore-A hardness did not vary as the magnetic field increased.

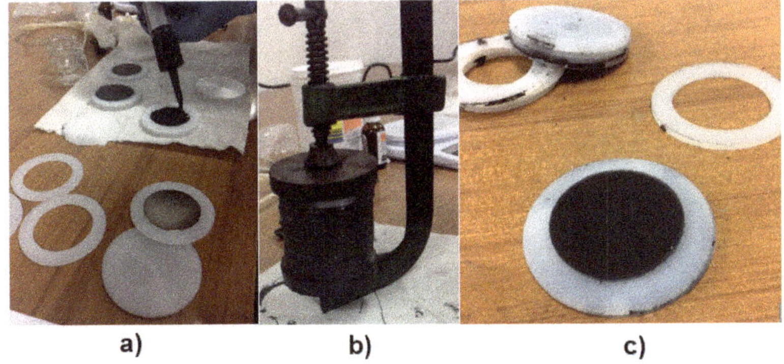

Figure 5. (**a**) Mold elements; (**b**) pouring phase; (**c**) MRE sample.

Table 1. MRE sample characteristics.

Matrix Material	Matrix Shore A Hardness	Compound Shore A Hardness	Matrix Density (g/mL)	Height (mm)	Diameter (mm)	Sample Density (g/mL)	Magnetic Field (mT)	Powder %
Silicone	10	33	1.08	6.0	50	2.59	257	25

4. Coil Characteristics

To characterize the material properties, the sample was tested by means of an experimental set-up containing a coil that produces a magnetic field whose strength lines are parallel to the particle columns of the MRE material. The tests were performed under different intensities of the magnetic field, which were obtained by adjusting the coil supply's current intensity.

To generate an adequate magnetic field, a coil and magnetic circuit were designed. In the design phase, to evaluate the magnetic flux that crosses the sample, an FEM model, developed in an Ansys Maxwell® environment, was adopted. The first configuration consisted of a cylindrical MRE element with a diameter and height of 50 and 10 mm, respectively; this element was inserted between two cylindrical steel elements (ϕ 60 × 10 mm). The coil had an inner diameter of 60 mm, an outer diameter of 108 mm, and a height of 100 mm. Inside the coil, there was a 100-mm high cylindrical steel core (Figure 6a). The saturation value of the magnetic field in the MRE was about 800 mT; to obtain it, 10,000 ampere-turns were required.

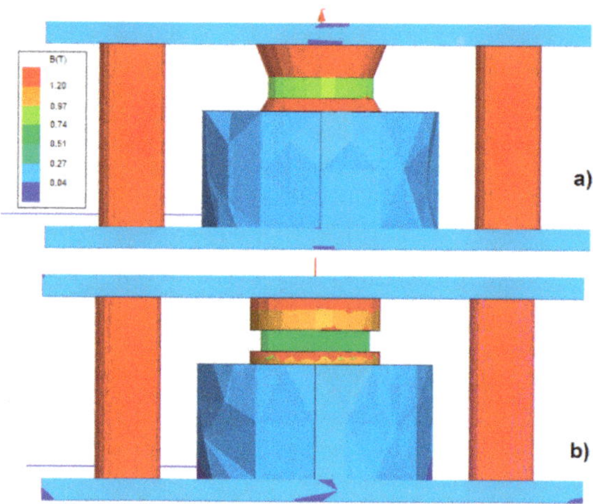

Figure 6. Magnetic field FEM model for 6000 ampere-turns: (**a**) cylindrical core; (**b**) conical core.

In the second configuration, to optimize the magnetic flux, the two cylindrical steel elements were replaced with two conical frustum elements with the following dimensions: bottom diameter: 60 mm; top diameter: 24 mm; height: 10 mm (Figure 6b). In this case, to reach the saturation value of the magnetic field, only 6000 ampere-turns were required.

The materials adopted for the FEM simulations were as follows: steel Aisi-1008 for the plates and core; copper for the coil; rubber and iron powder for the MRE pad. With respect to the second configuration (Figure 6b), considering a maximum current of 4 A, the coil required 1500 turns, was made of a 1.02-mm diameter cable (AWG17), and had an overall length equal to 310 m (resistance: 6.9 Ω; maximum dissipated power: 109 W).

5. MRE Characterization Tests

The realized samples were characterized by anisotropic chain-like structures, which were able to exhibit significant magneto-rheological effect if the magnetic field was applied in the direction of the particle chains that were parallel to the cylindrical samples' axes. Therefore, they were subjected to static forces acting along the samples' axes and to variable shear forces acting in the perpendicular direction. In both cases, the tests were performed under different intensities of the magnetic field, which were obtained by adjusting the coil supply's current intensity. The results of the two types of tests are described below:

(a) The cylindrical MRE samples were subjected to compression tests by means of a mechanical press to characterize the load–displacement trend and to identify the axial stiffness upon exposure to different values of the magnetic field. Another kind of axial test was then performed, compressing the samples between the two rigid plates of the mechanical press and keeping their distance constant; the axial load was detected for different values of the magnetic field intensity.

(b) The samples were tested under a constant compression load and a variable (harmonic) shear load. The tests were performed for different frequencies of the shear load and for different values of the magnetic intensity field.

5.1. Axial Characterization

The axial characterization was performed by assigning a continuously increasing and decreasing compression deformation to the MRE sample at the rate of about 0.01 mm/s. The test was repeated for different current intensities circulating in the coil windings. The axial force exerted by the screw press was detected by means of a load cell (Figure 7), while the corresponding deformation was detected by means of a centesimal comparator.

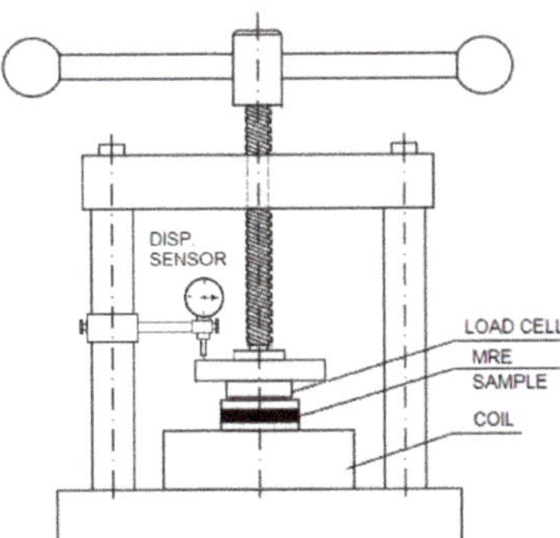

Figure 7. Scheme of the test rig used for axial characterization.

The first test was conducted with the material not activated (i.e., the coil not powered). The force–displacement trend (Figure 8) was almost linear (slightly hardening). At the beginning of the unloading phase (dashed line), the load exhibited a sharp reduction attributable to the material hysteresis. Figure 8 shows the test performed by feeding the coil at 2 A. There was an evident axial stiffening (about 65%), and in this case, at the beginning of the unloading phase, the diagram shows that there was a consistent force reduction.

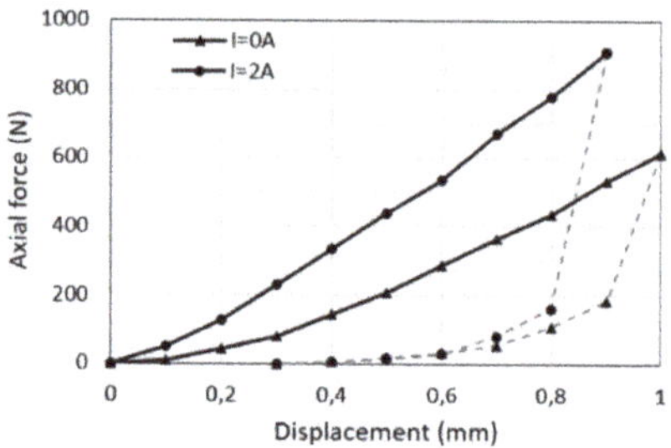

Figure 8. Force–displacement diagrams.

It is possible to compare the experimental results with the theoretical one obtained by means of the model presented in Section 2. For the prepared samples, the matrix modulus G_0 was equal to 0.3 MPa, and ϕ was equal to 0.25. Based on Equation (1), G = 0.66 MPa. Considering valid the relation for linear-elastic materials, Young's modulus is as follows: E = 2 G (1 + ν), where ν is the Poisson ratio. Considering the measured Poisson ratio (0.2) and the estimated G modulus (0.66 MPa), the following was determined: E = 1.57 MPa. In the case of rubber-like materials and for compressions limited to 10%, Hooke's law holds true [19].

The sample's theoretical axial stiffness can be evaluated as k = EA/h, where A is the sample's cross-section area, and h is the height. Figure 9 shows good agreement between the experimental and theoretical results for the two values of current I.

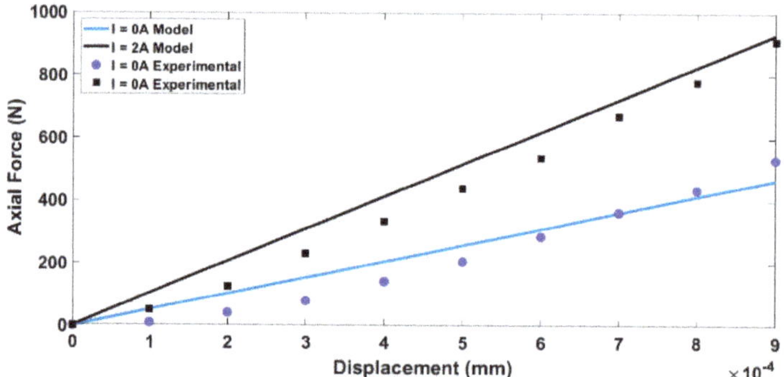

Figure 9. Comparison of the experimental and theoretical results.

Another kind of static test was performed on the sample; without feeding the coil, the sample was compressed between two plates. Keeping constant the distance between the plates, the axial force was then measured as the values of the coil current intensity increased. The results of this test, as reported in Figure 10, show that the axial force increased, with an almost linear trend, until the current intensity reached 6 A, with an axial load increment equal to about 63%; under greater current intensity, the axial load did not increase further (saturation).

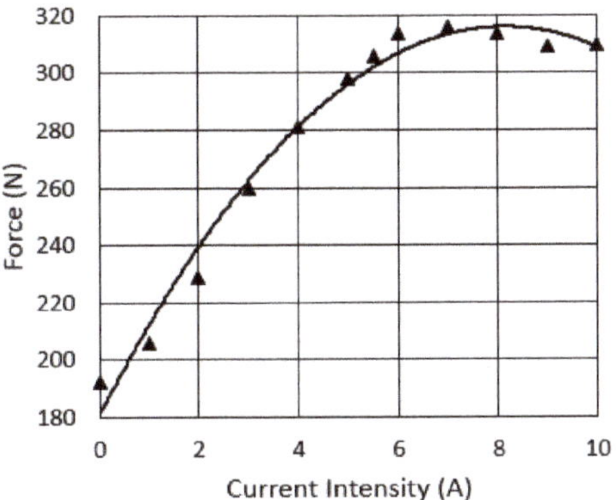

Figure 10. Axial force load vs. coil current intensity.

The delay of the load variation, with respect to the current variation, is shown in Figure 11 for both the load and unload phase; the test was performed by feeding the coil from 0 to 6 A and then returning to 0 A. The test results show that the MRE sample exhibited good reactivity during the increasing of the current intensity (loading), whereas, in the unloading phase, it reacted more slowly. In any case, the delay was compatible with a rapid and reversible control.

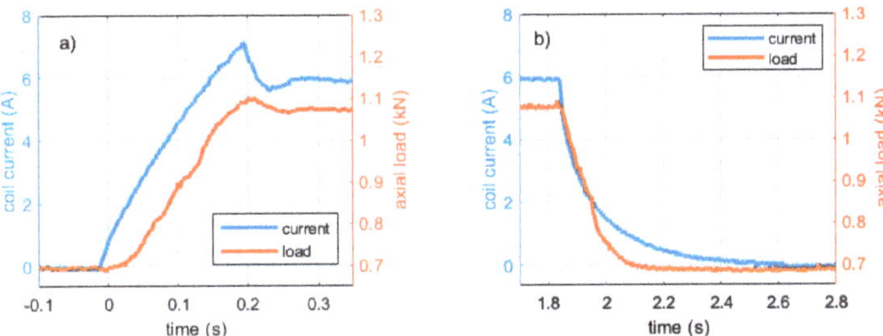

Figure 11. Axial force increments due to the change in current intensity (0–6 A): (**a**) current increased from 0 to 6 A; (**b**) current reduced from 6 to 0 A.

The magnetic field crossing the sample was measured by placing the gaussmeter probe between the sample and the plate. Figure 12 shows the relationship between the magnetic field H crossing the sample and the coil current intensity I. The curve was detected up until a current intensity of about 6 A, after which no stiffness variation occurred.

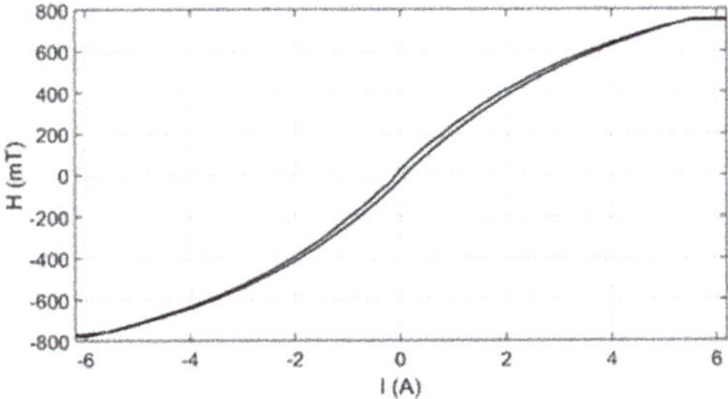

Figure 12. Magnetic field intensity vs. current intensity.

5.2. Shear Characterization

To characterize the MRE shear properties, the samples were tested under a constant compression load and variable shear excitation. The tests were conducted by placing two equal cylindrical samples (diameter: 50 mm; height: 6 mm) on opposite sides of a plastic platelet (Figure 13). The platelet, connected to an electro-dynamic shaker by means of a rod, provided a shear harmonic load to the samples. The samples were axially deformed by adjusting the distance between the two contact surfaces, which were composed of the magnetic core of a coil and a plate whose position could be adjusted to assign a desired axial deformation. The corresponding axial load ensured that the elements did not slip under the transverse load transmitted by the shaker.

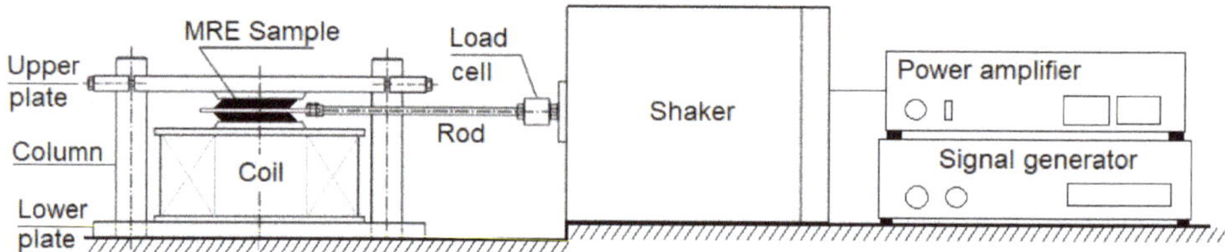

Figure 13. Scheme of the experimental set-up.

The test rig was equipped with an electromagnetic shaker (Bruel & Kjaer, mod. 4808), a coil able to generate 800 mT of magnetic field intensity, a load cell to measure the force exerted by the shaker (Dytran 6210S), an LVDT displacement sensor to measure the displacement of the shaker vibrating table (Inelta IGDL-5-k2455), and a force sensor resistor (FSR) that was placed between the specimen and the coil core to measure the axial load.

To highlight the influence of the forcing frequency and of the magnetic field intensity on the stiffness and damping properties of the MRE samples, several force–displacement diagrams were obtained for different operating conditions. Figure 14 shows the test results performed at the forcing frequencies of 5 and 10 Hz with two different values of coil current intensity (0 and 6 A).

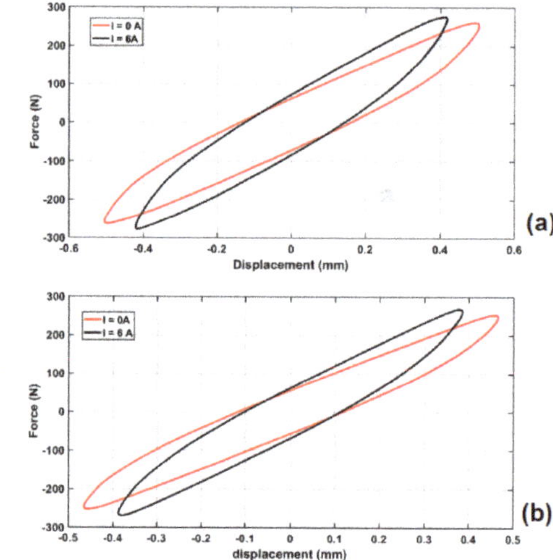

Figure 14. Shear tests: (**a**) f = 5 Hz; and (**b**) f = 10 Hz.

For each loop, the shear stiffness was estimated as the ratio between the maximum force and the maximum displacement, while the damping was evaluated through the cycle area by means of the expression: $\sigma = A/(\pi\omega X^2)$, where A is the cycle area, ω is the forcing circular frequency, and X is the motion amplitude.

The results of all the tests are summarized in the following diagrams. Figure 15 shows that the stiffness increased with the excitation frequency and with the intensity of the magnetic field. Furthermore, damping decreased with the excitation frequency (Figure 16) and increased as the coil current intensity increased.

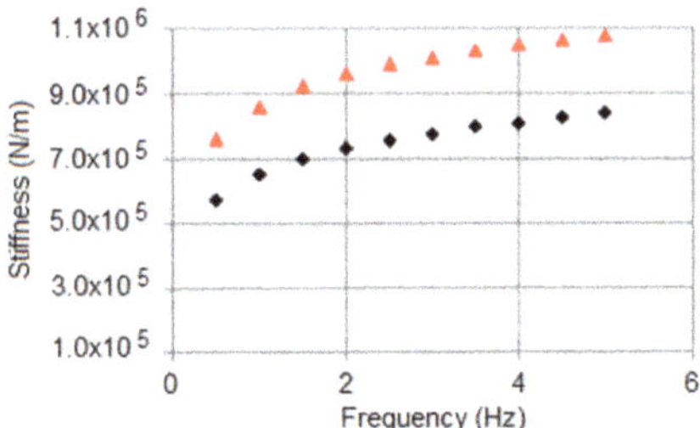

Figure 15. Stiffness vs. frequency.

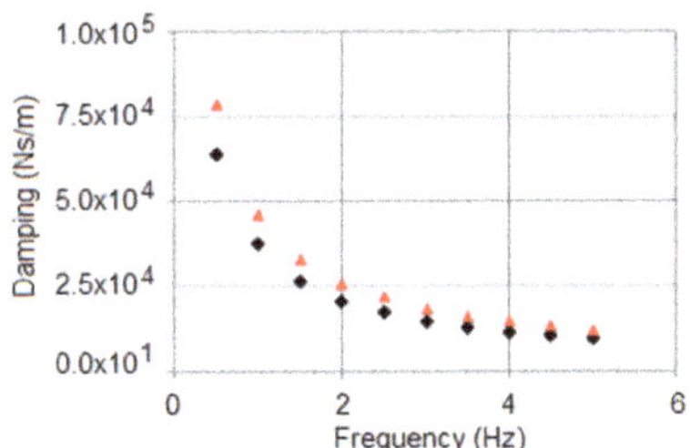

Figure 16. Damping vs. frequency.

The diagrams show that it was possible to control the shear stiffness and damping of the adopted MRE material at low frequencies, whereas at higher frequencies it was only possible to effectively control the stiffness.

5.3. Axial Preload Influence on Shear Performances

Finally, several tests were performed to investigate the preload influence on the sample's lateral stiffness. Force–displacements cycles (Figure 17) were detected under a forcing frequency of 2 Hz with two preload values (200 and 700 N) and two different values of coil current intensity (0 and 6 A). The graph shows that there was a stiffness increase in both cases; the percentage increase was greater in the case of the lower preload (+82%), while it was lower in the case of the higher preload (40%).

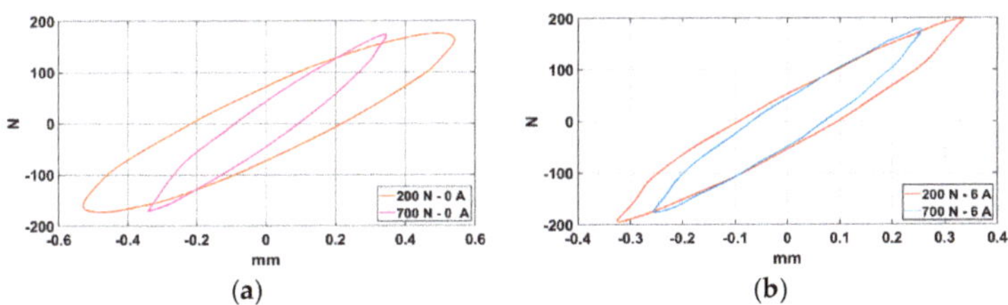

Figure 17. Sample tests for a forcing frequency of 2 Hz with a maximum horizontal force of 200 N: (**a**) current intensity of 0 A, shear stiffness: 330 N/mm at 200 N and 500 N/mm at 700 N; (**b**) current intensity of 6 A, shear stiffness: 600 N/mm at 200 N and 700 N/mm at 700 N.

6. MRE Dynamic Model

To analyze the dynamic behavior of the MRE element, we propose a theoretical model [20], whose scheme is reported in Figure 18.

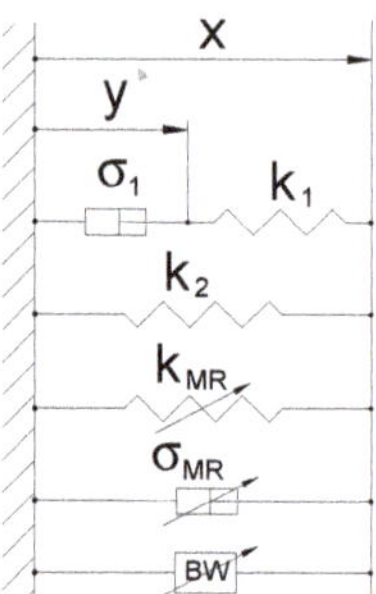

Figure 18. MRE visco-elastic model.

The proposed formulation uses the "standard linear model" to evaluate the linear visco-elastic behavior of the rubber. The hysteretic mechanism is well explained in [21]; the correspondent component can be modeled in several modes; some recent models having high computational efficiency are now available as, for example, that based on the shape function and memory mechanism [22] or the resistor-capacitor operator-based one [23]. In the following, the Bouc–Wen model [24] is adopted. The magneto-rheological effect is evaluated through the variable stiffness K_{mr} and the variable damping σ_{mr} and the restoring force is expressed by the following expressions:

$$F = k_1(x - y) + k_2 x + k_{MR} x + \sigma_{MR}\,\dot{x} + \alpha_{BW} w \tag{10}$$

$$k_1(x - y) = \sigma_1\left(\dot{x} - \dot{y}\right) \tag{11}$$

$$\dot{w} = \rho(\dot{x} - C|\dot{x}||w|^{n-1}w + (C - 1)\dot{x}|w|^n). \tag{12}$$

The adopted hysteretic model is characterized by the normalized variable w and by the following four parameters: C, ρ, n, and α_{BW}. Variable w assumes values confined within the range $[-1, 1]$ and is defined by means of Equation (12).

Figure 19 shows that there was a good agreement between the measured and simulated force–displacement cycles for the two different operating conditions.

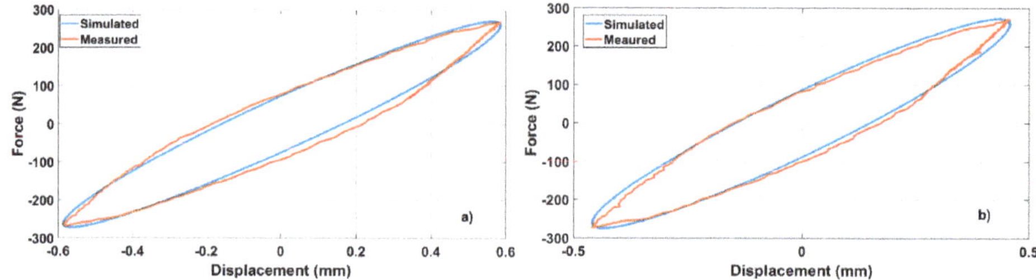

Figure 19. Simulated–measured cycle comparison: (**a**) 15 Hz frequency, 0 A current; (**b**) 15 Hz frequency, 6 A current.

7. Conclusions

The experimental tests conducted on the MRE samples, prepared with silicon elastomeric matrix and iron-carbonyl particles, were designed to evaluate their characteristics with respect to their potential application as vibration isolators.

A theoretical model determined the optimal volume fraction of the magnetic particles to maximize the magneto-rheological effect of the materials. Based on the theoretical results, several anisotropic MRE samples were prepared and tested using an experimental setup designed specifically for this study.

The experimental results demonstrated the wide variation in stiffness that can be achieved when such material is subjected to a magnetic field in both compression and shear tests; the material can, therefore, be used as a smart component of a semi-active vibration isolator.

Based on the test results, a visco-elastic model was developed to describe the sample's dynamic behavior. This model can be used to develop a control algorithm for a semi-active vibration isolation system and to simulate its behavior.

Author Contributions: All the authors participated in the: conceptualization, investigation, writing, and revision phases. G.D.M. took care of the experimental setup development and laboratory tests, R.B. developed the numerical models, and S.P. took care the results analysis. Funding was acquired by R.B. and G.D.M.

Funding: This research was funded by University of Naples Federico II under the project D.R.N.

Acknowledgments: The authors are grateful to Giuseppe Iovino and Gennaro Stingo for their collaboration during the setup construction and the execution of laboratory tests.

Conflicts of Interest: The authors declare no conflict of interest.

References

1. Sapouna, K.; Xiong, Y.P.; Shenoi, R.A. Dynamic mechanical properties of isotropic/anisotropic silicon magnetorheological elastomer composites. *Smart Mater. Struct.* **2017**, *26*, 115010. [CrossRef]
2. Li, Y.; Li, J.; Li, W.; Du, H. A state-of-the art review on magnetorheological elastomer devices. *Smart Mater. Struct.* **2014**, *23*, 123001. [CrossRef]
3. Li, J.; Li, Y.; Li, W.; Samali, B. *Development of Adaptive Seismic Isolators for Ultimate Seismic Protection of Civil Structures—Sensors and Smart Structures Technologies for Civil, Mechanical, and Aerospace Systems*; SPIE: San Diego, CA, USA, 2013.
4. Behrooz, M.; Wang, X.; Gordaninejad, F. Modelling of a new semi-active/passive magnetorheological elastomer isolator. *Smart Mater. Struct.* **2014**, *23*, 045013. [CrossRef]
5. Di Massa, G.; Pagano, S.; Strano, S. Cabinet and shelter vibration isolation: Numerical and experimental investigation. *Eng. Lett.* **2014**, *22*, 149–157.
6. Gong, X.L.; Zhang, X.Z.; Zhang, P.Q. Fabrication and characterization of isotropic magnetorheological elastomers. *Polym. Test.* **2005**, *24*, 669–676. [CrossRef]
7. Bica, I. The influence of the magnetic field on the elastic properties off anisotropic magnetorheological elastomers. *J. Ind. Eng. Chem.* **2012**, *18*, 1666–1669. [CrossRef]
8. Brancati, R.; di Massa, G.; Pagano, S.; Strano, S. A Seismic Isolation System for Lightweight Structures based on MRE Devices. In Proceedings of the World Congress on Engineering 2015, London, UK, 1–3 July 2015.
9. Brancati, R.; di Massa, G.; di Vaio, M.; Pagano, S.; Santini, S. Experimental investigation on magneto-rheological elastomers. In Proceedings of the Second International Conference of IFToMM Italy (IFIT2018), Cassino, Italy, 29–30 November 2018.
10. Ubaidillah, U.; Sutrisno, J.; Purwanto, A.; Mazlan, S.A. Recent progress on magnetorheological solids: Materials, fabrication, testing and applications. *Adv. Eng. Mater.* **2015**, *17*, 563–597. [CrossRef]
11. Kallio, M. The Elastic and Damping Properties of Magnetorheological Elastomers. Ph.D. Thesis, VTT Technical Research Centre of Finland, Espoo, Finland, 2005.
12. Guth, E.J. Theory of filler reinforcement. *J. Appl. Phys.* **1945**, *16*, 20–25. [CrossRef]
13. Jolly, M.R.; Carlson, J.D.; Munoz, B.C. Model of the behaviour of magnetorheological materials. *Smart Mater. Struct.* **1996**, *5*, 607–614. [CrossRef]
14. Rosensweig, R.E. *Ferro Hydro Dynamics*; Cambridge University Press: Cambridge, UK, 1985.

15. Jiles, D. *Introduction to Magnetism and Magnetic Materials*, 3rd ed.; C.R.C. Press: Boca Raton, FL, USA, 2009.

16. Wu, J.; Gong, X.; Chen, L.; Xia, H.; Hu, Z. Preparation and characterization of isotropic polyurethane magnetorheological elastomer through in situ polymerization. *J. Appl. Polym. Sci.* **2009**, *114*, 901–910. [CrossRef]

17. Yang, C.C.Y.; Fu, J.; Yu, M.; Zheng, X.; Ju, B.X. A new magnetorheological elastomer isolator in shear–compression mixed mode. *J. Intel. Mat. Syst. Struct.* **2015**, *26*, 1290–1300. [CrossRef]

18. Brancati, R.; di Massa, G.; Pagano, S. A vibration isolator based on magneto-rheological elastomer. In *Advances in Italian Mechanism Science, Proceedings of the First International Conference of IFToMM Italy (IFIT2016), Vicenza, Italy, 1–2 December 2016*; Springer: Basel, Switzerland, 2017; pp. 483–490.

19. Kukla, M.; Gòrecki, J.; Malujda, I.; Talaśka, K. The determination of mechanical properties of magnetorheological elastomers (MREs). *Procedia Eng.* **2017**, *177*, 324–330. [CrossRef]

20. Brancati, R.; di Massa, G.; Pagano, S.; Santini, S. A magneto rheological elastomer vibration isolator for Lightweight STRUCTURES. *Meccanica* **2019**, *54*, 333–349. [CrossRef]

21. Bai, X.X.X.; Chen, P. On the hysteresis mechanism of magnetorheological fluids. *Front. Mater. Sci.* **2019**, *6*, 36. [CrossRef]

22. Chen, P.; Bai, X.X.; Qian, L.J.; Choi, S.B. An approach for hysteresis modeling based on shape function and memory mechanism. *IEEE-ASME Trans. Mech.* **2018**, *23*, 1270–1278. [CrossRef]

23. Bai, X.X.; Cai, F.L.; Chen, P. Resistor-capacitor (RC) operator-based hysteresis model for magnetorheological (MR) dampers. *Mech. Syst. Signal. Process.* **2019**, *117*, 157–169. [CrossRef]

24. Ismail, M.; Ikhouane, F.; Rodellar, J. The hysteresis bouc-wen model, a survey. *Arch. Comput. Methods Eng.* **2009**, *16*, 161–188. [CrossRef]

Article

Experimental Characterization of the Coupling Stage of a Two-Stage Planetary Gearbox in Variable Operational Conditions

Claudia Aide González-Cruz [1,*] and Marco Ceccarelli [2]

[1] Universidad Autónoma de Querétaro, Dirección de Investigación y Posgrado de la Facultad de Ingeniería, Querétaro QRO 76010, Mexico

[2] University of Rome Tor Vergata, Department of Industrial Engineering, Laboratory of Robot Mechatronics, 00133 Rome, Italy; marco.ceccarelli@uniroma2.it

* Correspondence: claudia.aide.gonzalez@gmail.com; Tel.: +52-1-442-475-4235

Received: 16 April 2019; Accepted: 17 June 2019; Published: 20 June 2019

Abstract: This paper presents an experimental characterization of a two-stage planetary gearbox (TSPG) designed at the Laboratory of Robot Mechatronics (LARM2) in the University of Rome Tor Vergata. The TSPG operates differentially as function of the attached load and the internal friction forces caused for the contact between gears. Experiments under varying load conditions are developed in order to analyze the usefulness of the gearbox to avoid excessive torques on its internal elements. The analysis of the dynamic torques is presented as an indicator of stability in the gearbox operation. The results show that the actuation of the second operation stage reduces the torques 57% in the output shaft and 65% in the input shaft. The efficiency of the gearbox is estimated as 40% in presence of high internal friction forces.

Keywords: two-stage planetary gearbox; varying load; dynamic torque; efficiency; rotating machinery

1. Introduction

The planetary gearbox is one of the most used mechanical transmission designs due to the advantages that it has over the gearbox with simple transmission, i.e., high torque, high transmission rate, high speed rate, compact volume design, to mention a few. However, planetary gearboxes are characterized by a complex kinematics and many internal and external excitation sources, such as varying mesh and bearing stiffness, assembly errors, manufacturing errors and varying operation conditions of speed and load. These excitation sources may increase the modal interaction in non-linear regimes and produce chaotic behavior, nonlinear jump, resonance and bifurcation. Therefore, there is a constant interest for modelling the dynamics of planetary gearboxes.

Modal analysis is a widely used technique to calculate the frequency response of a system and thus, determine the natural frequencies, modal damping, and the mode shapes. Experimentally, it is based on impact test and an excited mechanical structure by means of a hammer and a shaker, respectively. Nevertheless, it has been proved that experimental modal analysis suffers from missing modes because some modes cannot be excited [1]. Numerous research works on condition monitoring and fault diagnosis have been accomplished with the aim of preserve planetary gearboxes in healthy operating conditions. Many of them have been focused on the analysis of the stationary operation process [2–4]. However, stationary conditions may hide important transient features about the machinery condition that can be exposed only during non-stationary operation, especially during the run-up and run-down process [5,6].

Flexible and rigid body models have been developed by researchers with the aim of reproducing the dynamic behavior of planetary gearboxes, as well as developing fault diagnosis techniques.

The effect of the elasticity of the ring gear on the modal properties and time varying mesh stiffness of a planetary gear system has been studied through the variation of design parameters, such as the number of planets, ring-bending stiffness and sliding friction coefficient in the dynamic model [7]. The modulation sidebands induced by a single planet gear mesh and multiple planet gears meshes on the dynamic response of a planetary gear have been studied using models based on signal transfer path functions and measuring-direction projection functions of the gear mesh force. The analysis of the significance level of meshing frequency modulation can reveal important dynamic characteristics for failure identification [8,9] and even to distinguish whether the planetary gear train is assembled with a floating sun gear, or whether the planetary gear train without a floating sun gear exists with some distributed defects [10].

The modulation phenomenon caused by the time-varying gear meshing and position of the planet gears is widely studied. Vibration signals at the transducer location are expressed by the sum of all vibrations induced from each component and influenced by the time-varying path [11]. The geometry of the tooth profile has an important role in the gear meshing in planetary gear trains. Models based on the exact involute gear geometry have been proposed as a tool for the determination of tolerances for the movable area and assemblability of the sun gear, the backlash and other components of the gear train [12]. It also has been demonstrated that the tooth wear has an important effect on the bending stiffness, shear stiffness, and axial compressive stiffness because they depend of the tooth profile and tooth thickness, and the time-varying mesh stiffness decreases according to the tooth wear depth [13].

Accelerometers are the transducers mostly used to capture the dynamic behavior of rotating machinery. The location of the accelerometers on the gearbox structure plays an important role, especially in planetary gearboxes, which have multiple planets producing similar vibrations when the planets pass through the fixed sensor [14]. On the other hand, even when there are experimental works that do not require angular information [15,16], tachometers are necessary to measure important variable speed conditions. Other transducers widely used are extensometers and thermocouples, since they allow to evaluate the performance efficiency and the energy loss during the torque transmission [17].

In this context, this paper presents an experimental analysis of feasibility for a two-stage planetary gearbox (TSPG) prototype at the Laboratory of Robot Mechatronics (LARM2) in the University of Rome Tor Vergata, with the aim of validate its functionality and quantify its efficiency and dynamic behavior. Experiments are carried out in a test bed which allows to develop non-stationary operating conditions by means of varying the load torque and measuring multiple variables.

2. Gearbox Design

A schematic of the planetary gearbox under study is shown in Figure 1 [18]. It consists of two-stage planetary gear trains in parallel with floating sun gears. Each planetary train comprises one carrier $C^{(n)}$, two planet gears $P^{(n)}$, one sun gear $S^{(n)}$ and one internal ring gear $R^{(n)}$; the superscript n indicates the corresponding planetary train, $n = 1$ for the input train and $n = 2$ for the output train. Note that since the sun gears are fixed on the same shaft and the ring gears are on the same frame, each couple of gears actuates as a rigid body, $S^{(1,2)}$ and $R^{(1,2)}$. The input torque is applied to the sun gear $S^{(1)}$ and ring gears $R^{(1)}$ from the input carrier $C^{(1)}$ by means of the planet gears $P^{(1)}$. Since sun gears and ring gears act like a rigid body, the torque is transmitted to the planet gears $P^{(2)}$ to the output carrier $C^{(2)}$.

The TSPG mechanism operates differentially as function of the load torque and the internal friction forces that are caused by the contact between gears, Figure 2. The first operation stage (FOS) of the planetary gearbox runs when the torque caused for the attached load is low, Figure 2a. In this operating condition, the friction forces between gears cause a self-locking phenomenon that locks the planet gears to the sun and ring gears. As consequence, they do not rotate over their own axes, $\omega_P^{(1)} = \omega_P^{(2)} = 0$. Consequently, the closed gear chain $P^{(1)}$-$R^{(1)}$-$R^{(2)}$-$P^{(2)}$-$S^{(2)}$-$S^{(1)}$ behaves as a unique rigid body which rotates about the axis of the two sun gears. Thus, the movement of the input carrier is directly transmitted to the output carrier, giving $\omega_C^{(2)} = \omega_C^{(1)}$. The second operating stage (SOS) of the planetary gearbox runs when the load torque increases, Figure 2b, and causes that the output carrier remains locked in a fixed position with $\omega_C^{(2)} = 0$. Meanwhile the friction forces between gears are overcome and the planet gears start to rotate around their own axes with $\omega_P^{(1),(2)} \neq 0$. Consequently, the frame of the ring gears becomes the output link rotating in opposite direction to the input carrier.

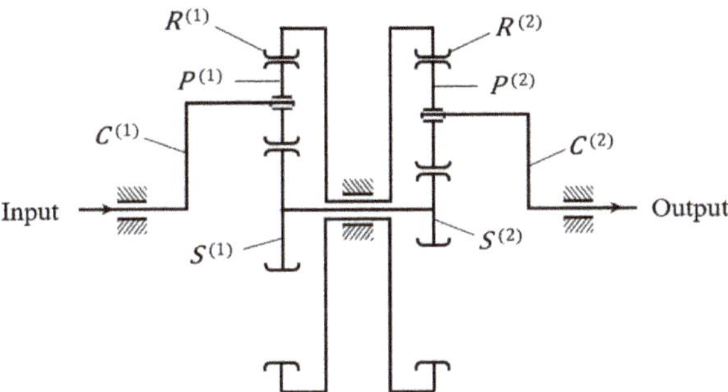

Figure 1. Structure diagram of the two-stage planetary gearbox ($C^{(n)}$ - carrier, $S^{(n)}$ - sun gear, $P^{(n)}$ - planet gears, $R^{(n)}$ is ring gear; with $n = 1$ for the input planetary train and $n = 2$ for the output planetary train).

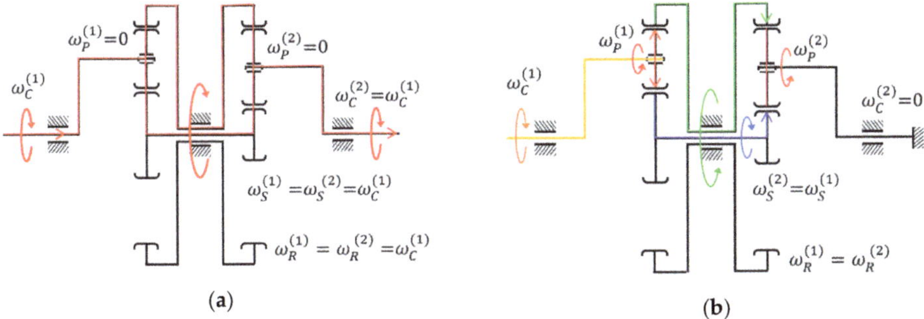

(**a**)

(**b**)

Figure 2. Operating scheme of the two-stage planetary gearbox: (**a**) first stage; (**b**) second stage.

The operational characteristics of the gearbox enable the self-exchanging of the output link under variable load conditions. Thus, when the load increases suddenly the actuation of the SOS helps to reduce the stress in the internal gearbox elements.

The prototype of the TSPG is shown in Figure 3. All gears are spur gears with modules equal to 1 mm and pressure angle equal to 20°. The design parameters of the gears are listed in Table 1.

The gear ratio $u^{(n)}$, for the input ($n = 1$) and output ($n = 2$) planetary gear trains in Figure 1, can be expressed as function of the teeth number of the gears N and the angular velocity ω of each element as follows:

$$u^{(1)} = -\frac{N_R^{(1)}}{N_S^{(1)}} = \frac{\omega_S^{(1)} - \omega_C^{(1)}}{\omega_R^{(1)} - \omega_C^{(1)}} \tag{1}$$

$$u^{(2)} = -\frac{N_R^{(2)}}{N_S^{(2)}} = \frac{\omega_S^{(2)} - \omega_C^{(2)}}{\omega_R^{(2)} - \omega_C^{(2)}} \tag{2}$$

By solving Equations (1) and (2), the angular velocity of the ring gear and the sun gear are given by

$$\omega_R = \frac{\left(u^{(1)} - 1\right)\omega_C^{(1)} - \left(u^{(2)} - 1\right)\omega_C^{(2)}}{u^{(1)} - u^{(2)}} \tag{3}$$

$$\omega_S = \left(\omega_R^{(2)} - \omega_C^{(1)} \right) u^{(1)} + \omega_C^{(1)} \tag{4}$$

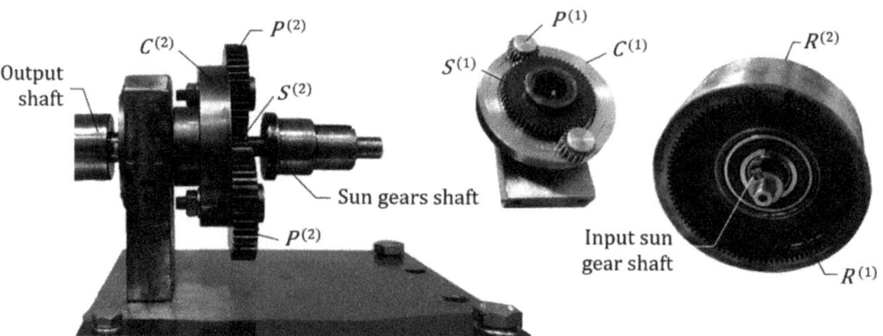

Figure 3. Prototype of the two-stage planetary gearbox under study ($C^{(n)}$- carrier, $S^{(n)}$- sun gear, $P^{(n)}$- planet gears, $R^{(n)}$ - ring gear; with $n = 1$ for the input gear train and $n = 2$ for the output gear train).

Table 1. Design parameters of the Laboratory of Robot Mechatronics (LARM2) planetary gearbox prototype.

Element	Number of Teeth, N	Pitch Radius, r [mm]
Input gear set		
Sun gear	60	30
Planet gear	20	10
Ring gear	100	50
Carrier		40
Output gear set		
Sun gear	12	6
Planet gear	44	22
Ring gear	100	50
Carrier		28
Module	1	
Pressure angle	20°	

On the other hand, the torques transmitted for the planetary gearbox can be obtained from a static equilibrium analysis as shown in Figure 4. Where, $F_m^{(n)}$ and $r_m^{(n)}$ are the transmitted force and the radius of the element m, respectively. Thus, the torques transmitted by the input and output carriers are given as:

$$T_C^{(1)} = F_C^{(1)} r_C^{(1)} \tag{5}$$

$$T_C^{(2)} = F_C^{(2)} r_C^{(2)} \tag{6}$$

Finally, the efficiency of the planetary gear train can be obtained by:

$$\eta = \frac{W^{(2)}}{W^{(1)}} \tag{7}$$

where, $W^{(n)}$ is the mechanical power of the input and output gear trains, $n = 1, 2$ respectively, given by:

$$W^{(n)} = M_C^{(n)} \omega_C^{(n)} \tag{8}$$

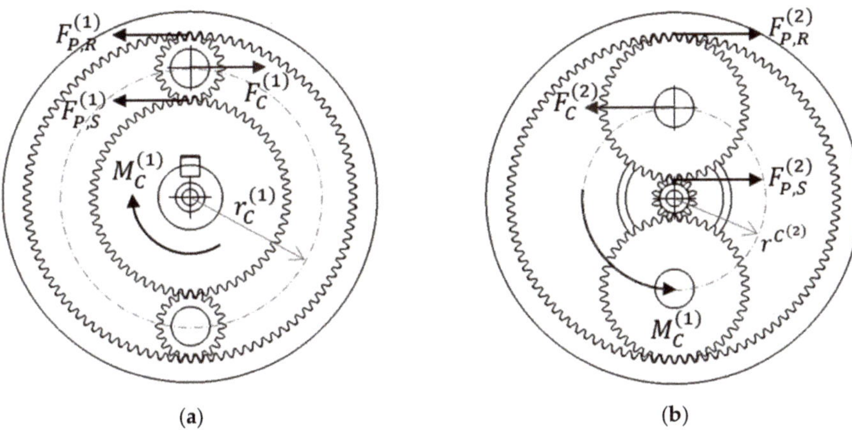

Figure 4. A free body diagram for static equilibrium: (**a**) input gear set; (**b**) output gear set.

3. Test-Bed Design

A new test-bed configuration was designed in order to characterize the dynamic performance of the two-stage planetary gearbox under study. However, the design characteristics of the test-bed enable the development of test of other rotating machinery with similar operating requirements. The test-bed was projected with the aim to test the full operation condition of the planetary gearbox prototype, meanwhile the most representative dynamic variables are measured.

Figure 5 shows the schematic diagram and the experimental setup of the test-bed. It consists of an AC motor that is coupled to a planetary gearbox prototype to drive the attached load. The operating speed of the motor is controlled by means of a frequency inverter, while the load torque is variated by means of a mechanical break, which actuates directly on the load surface.

The test-bed is instrumented to measure the most representative variables for rotating machinery. Two biaxial accelerometers are placed on the bearings that support the input and output shafts of the gearbox, respectively, to measure the radial and axial vibrations. One uniaxial accelerometer is placed on the bearing that supports the output shaft near the load to measure the radial vibrations; as indicated in Figure 5. Two dynamic torque sensors are placed in the input and output shafts of the gearbox, respectively, to measure the torque variation during the exchange between the FOS and SOS. Finally, one encoder is placed in the output shaft to measure the angular velocity.

The data acquisition system consists of a signal conditioning stage to amplify the torque signals, two DAQ boards from National Instruments and one DAQ board from Phidgets Inc. The development of the tests is supervised by means of a graphical user interface developed in LabView. Furthermore, from the user interface it is possible to control the acquisition of data, the sampling frequency and the data storage in the PC. The list of the equipment installed in the test-bed is given in Table 2.

The test-bed enables the development of different kind of tests, such as: (1) run-up and run-down speed profile tests, (2) constant operating speed and constant load, (3) varying operating speed and varying load. In this work, the experience with the TSPG prototype is reported at constant operational speed and varying load.

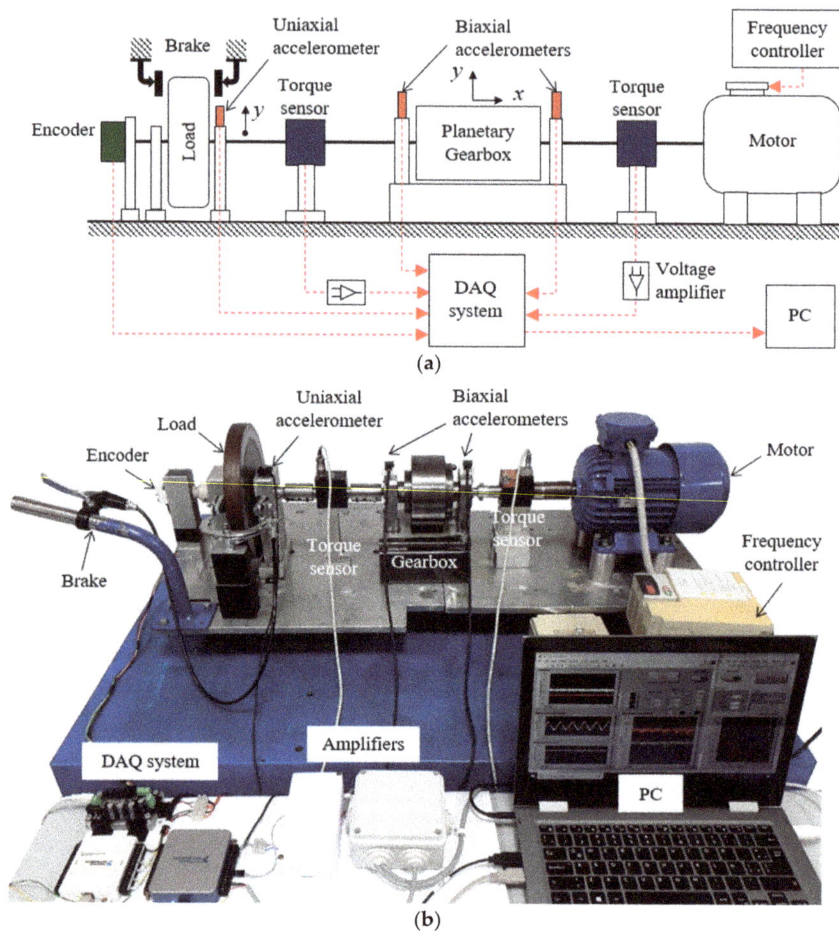

Figure 5. Test-bed for experiments on rotating machinery: (**a**) schematic diagram; (**b**) experimental setup.

Table 2. Equipment used in the test-bed.

Equipment	Model	Quantity
AC Motor	Cantoni Sh-80-2B	1
Frequency inverter	Mitsubishi FR-E500	1
Accelerometers	ADXL-321J	2
	Kistler 8305B	1
Torque sensors	CD1050	1
	CD1095	1
Encoder	ENC1J-D16-L00128L	1
Data acquisition system	NI USB 6009	1
	NI USB 6001	1
	Phidget 1065	1

4. Experimental Results

Experiments are carried out at constant speed and varying load. The results for the two different speed conditions ω_n= 120 rpm and ω_n= 210 rpm are presented in this work; where ω_n indicates the nominal speed that the motor drive reaches. The load condition is variated during the experiments from low to high by means of brake system actuation. The term low load refers to the load condition produced by the single flywheel, while the term high load refers to the load condition produced when the brake actuates on the flywheel surface.

Each experiment is conducted as follows: (1) the system is started up with low load condition; therefore, the FOS of the TSPG is actuated; (2) when the output shaft reaches the nominal speed ω_n, the load is increased; consequently, the SOS of the TSPG is actuated; (3) when the system reaches the nominal operating conditions, the load is decreased; thus, the FOS of the gearbox is reactivated. Experimental results are presented below in order to analyze the dynamics of the gearbox during the operational exchange between FOS and SOS.

Figure 6 shows the experimental results for the test at $\omega_n = 120$ rpm. It can be seen that the amplitude of the torques and vibrations varies as a response of the load variation. The dynamic torques in the input and output shafts are shown in Figure 6a. It can be seen that both input and output torques have similar behavior. The torque amplitudes increase as the load increases and they decrease as the load decreases. However, the input shaft torque is always higher than the output shaft torque. This is because the torque in the input shaft is imposed by the torque in the output.

Figure 6a shows that the input and output shaft torques increase during the start-up of the system until they reach the maximum values $T_{in} = 1.38$ Nm and $T_{out} = 0.54$ Nm, respectively. This increment is produced by the inertial mass of the coupled system gearbox-load. Once the inertial mass is overcome, the coupled system starts to rotate and the torques decrease until they reach nominal values under low load condition, $T_{in} = 0.94$ Nm and $T_{out} = 0.30$ Nm.

The high load condition in the output shaft is generated by means of the actuation of the brake system, Figure 6b. As consequence, a sudden increment in the torques arises, with peak values $T_{in} = 2.17$ Nm and $T_{out} = 0.97$ Nm. Then, the SOS of the gearbox is self-actuated and the torques decrease until they reach the nominal values under high load condition, $T_{in} = 1.41$ Nm and $T_{out} = 0.56$ Nm. It is important to note that the nominal values of the torques under high load condition are higher than those under low load condition. This is because the higher the load, the higher the input torque to drive it.

The usefulness of the SOS in the gearbox design can be seen as the reduction of the torque in the gearbox elements once the system achieves stable operation under high load condition. Thus, in this work, the torque reduction is estimated as the difference between the peak value of the torque when the load increases and its nominal value after the system achieves the stable operation condition. From the experimental data at $\omega_n = 120$ rpm, it is estimated that the actuation of the SOS of the gearbox enables the reduction of torque 57% in the output shaft and 65% in the input shaft.

After the stable operation condition with the SOS is achieved, the load torque is decreased and the FOS of the gearbox is reactivated. It can be seen in Figure 6a that the torques decrease smoothly until they reach again the nominal values under low load condition. A disturbance in the torque signals appears when the output shaft reaches the nominal speed $\omega_n = 120$ rpm. This disturbance is produced by the self-locking phenomenon of the gears that occurs when the system achieves the nominal operation conditions for the FOS.

The small disturbances in the torque signals caused for the self-locking phenomenon of the gears when the system reach the nominal operation conditions for the FOS can be seen again in Figure 7a.

The angular speed of the output shaft is shown in Figure 6b. It can be seen that during the operation of the FOS, the rotation of the input shaft is directly transmitted to the output shaft, while during the operation of the SOS, the output shaft remains braked. It is important to note that the variations in the angular velocity measurements are due to the quality of the encoder used during the experiments.

The radial vibrations on the input and output shafts during the test at $\omega_n = 120$ rpm are shown in Figure 6c,d, respectively. The vibrations at the beginning of the test are generated due to the overcoming of the inertial mass gearbox-load. Furthermore, it can be seen that the vibration amplitudes increase suddenly when the SOS is actuated and they decrease as the system reaches nominal operational conditions. It is assumed that the higher amplitudes of vibration in the input shaft than those in the output shaft are because the input carrier is rotating while the output carrier remains braked in a fixed position.

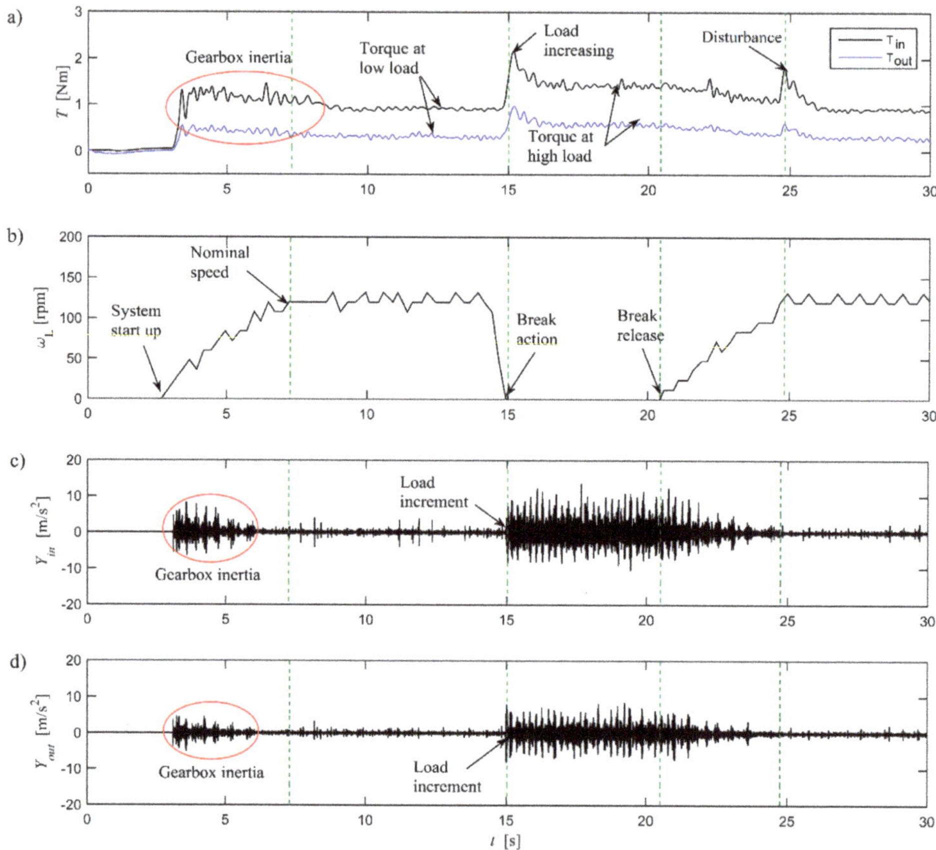

Figure 6. Waveforms at 120 rpm operation speed: (**a**) dynamic torques; (**b**) angular speed of the output shaft; (**c**) radial vibrations on the input shaft; (**d**) radial vibrations on the output shaft.

The experimental results for the test at $\omega_n = 210$ rpm are shown in Figure 7. Similar to the previous experimental case, it can be seen that the amplitude of the torques and vibrations varies in response to the varying load condition. Their amplitudes are always higher in the input shaft than in the output shaft.

The torques in the input and output shafts are shown in Figure 7a. It can be seen that their dynamic behavior is similar to that of the previous test. The maximum amplitude of the torque caused for the inertial mass gearbox-load during the system star-up are $T_{in} = 1.75$ Nm and $T_{out} = 0.77$ Nm. Once the inertial mass of the coupled system gearbox-load is overcome, the FOS of the gearbox starts to operate. The nominal torque values for the input and output shafts under low load condition are $T_{in} = 0.89$ Nm and $T_{out} = 0.34$ Nm, respectively. The SOS of the gearbox is actuated when the load increases. The maximum peak values of torques under high load condition are $T_{in} = 2.40$ Nm and $T_{out} = 1.20$ Nm. The torque amplitudes decrease as the system achieves the nominal operation conditions for the SOS. The nominal values of torques under high load condition are $T_{in} = 1.83$ Nm and $T_{out} = 0.76$ Nm.

The reduction of torques during the operation of the SOS of the gearbox is estimated as described before. The reduction of torques for this test is 63% in the output shaft and 76% in the input shaft. The disturbance in the torque signals caused for the self-locking phenomenon of the gears that occurs when the system achieves the nominal operation conditions for the FOS can be seen again in Figure 7a.

The angular speed of the output shaft is shown in Figure 7b and the radial vibrations on the input and output shafts are shown in Figure 7c,d, respectively. It can be seen that the vibrations increase during the overcoming of the inertial mass of the coupled system gearbox-load at the beginning of the test. Furthermore, it can be seen that the amplitude of vibrations increases suddenly when the SOS of the gearbox is actuated and also it increases at higher operational speed.

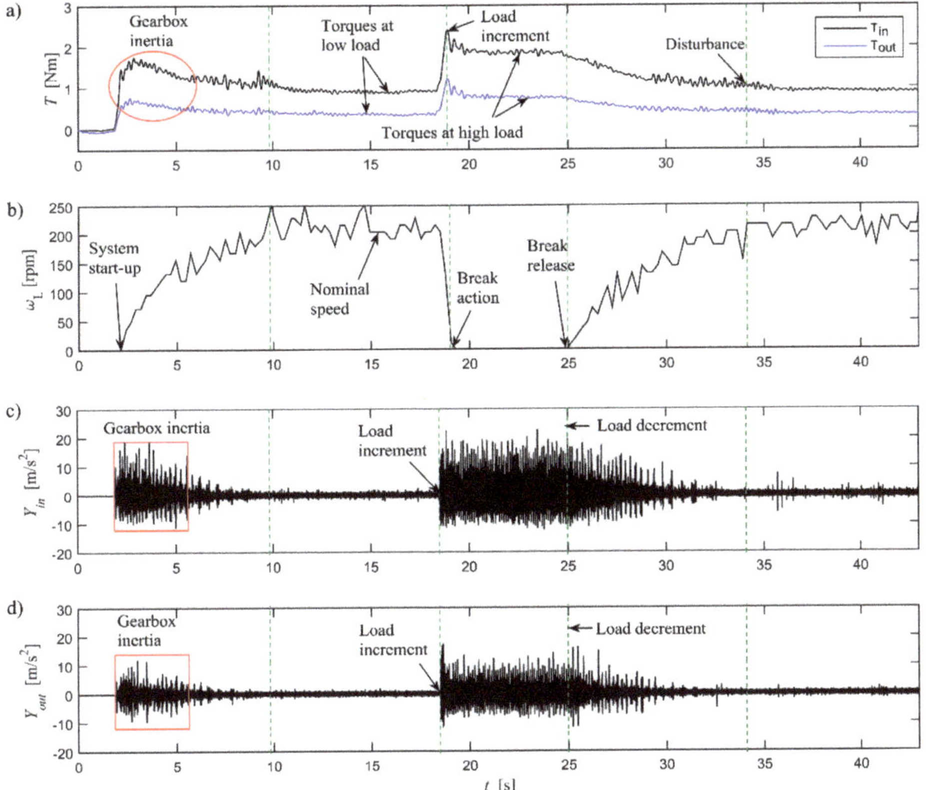

Figure 7. Waveforms at 210 rpm operation speed: (**a**) dynamic torques; (**b**) angular speed of the output shaft; (**c**) radial vibrations on the input shaft; (**d**) radial vibrations on the output shaft.

5. Discussion

An estimation of parameters such as inertial mass, friction torque and efficiency of the gearbox can be made from experimental data. In this work, these parameters are estimated from the torque signals as follows: (1) the inertial mass M_I of the coupled system gearbox-load is estimated from the input torque data as the difference between the maximum value during the start-up process of the system and the nominal value at low load condition; (b) the friction torque of the internal elements T_f is estimated as the difference between the peak value when the load increases and the nominal value at high load condition; (c) the efficiency η is estimated as the rate between the input and output torques.

These parameters are estimated from the experimental data at ω_n =120 rpm. The estimated values are $M_I = 0.89$ Nm, $T_f = 1.23$ Nm and $\eta \approx 40\%$.

The low transmission efficiency is expected since the gearbox housing is missing in the prototype. This fact generates a low lubrication condition for the internal gearbox elements and, consequently, it produces an increment in the internal friction forces. The amplitude of the torques is higher during the SOS than during the FOS because the higher load in the output. Nevertheless, the experimental results prove the usefulness of the SOS in the gearbox design, since it is self-actuated when there is a sudden increment in the load conditions. It can be seen in Figures 6a and 7a that the torque increases suddenly when the load torque increases, but the torque amplitude decreases as the SOS achieves the nominal operation conditions. Thus, the actuation of the SOS reduces stress on the gearbox elements.

6. Conclusions

A new configuration of the test bed is presented in order to evaluate the dynamic behavior of a two-stage planetary gearbox (TSPG). The conceptual design of the test-bed enables the development of experiments on rotating machinery with similar operating requirements. In this work, experimental tests at constant operating speed and varying load are carried out in order to test the operation of the TSPG designed at LARM2.

The TSPG mechanism consists of a two-stage planetary gearbox in parallel with floating sun gears. The mechanism operates differentially as a function of the load torque and the internal friction forces that are caused by the contact between gears. The first operation condition (FOS) is self-actuated under low load conditions. The second operation stage (SOS) is self-actuated under high load conditions. The characteristics of the gearbox enable the self-exchanging of the output link under variable load conditions. The SOS enables the ring gear as the output link when the load in the output shaft increases suddenly. As a consequence, the stress in the internal gearbox elements is reduced under high load operation conditions.

The experimental results demonstrate the usefulness of the gearbox design since the SOS is self-actuated when the load in the output shaft increases suddenly. The high load condition of the gearbox produces a sudden increment in the amplitude of the torques. However, the torques decrease as the SOS achieves the nominal operation conditions. Thus, the actuation of the SOS reduces the stress on the gearbox elements. The results show that the actuation of the SOS reduces the torque 57% in the output shaft and 65% in the input shaft during the test at 120 rpm and 63% and 76%, respectively, during the test at 210 rpm. The efficiency, inertial mass of the coupled system gearbox-load and the friction torque in the internal gearbox components are estimated from the torque signals. It is found that the efficiency of the gearbox is $\eta \approx 40\%$. This value is expected to be low since the lubrication of the gears is absent due to the missing housing of the gearbox. The inertial mass of the coupled system gearbox-load is estimated as $M_I = 0.89$ Nm and the friction torque generated by in the internal gearbox elements is $T_f = 1.23$ Nm.

Future work is planned for the modelling of the TSPG mechanism and the identification of parameters from vibration signals, such as the design modifications in order to improve the efficiency and to use the torque transmitted by the ring gear during the SOS.

Author Contributions: Conceptualization, C.A.G.-C. and M.C.; methodology, C.A.G.-C. and M.C.; software, C.A.G.-C.; validation, C.A.G.-C. and M.C.; formal analysis, C.A.G.-C.; investigation, C.A.G.-C. and M.C; resources, M.C.; data curation, C.A.G.-C.; writing—original draft preparation, C.A.G.-C.; writing—review and editing, M.C.; visualization, C.A.G.-C.; supervision, M.C.; project administration, C.A.G.-C. and M.C.; funding acquisition, M.C.

Funding: This research received no external funding.

Acknowledgments: The first author would like to gratefully acknowledge the Mexican Government Foundation CONACYT for the fellowship to pursue her postdoctoral studies at LARM2 in the academic year 2018-2019. The authors wish to thank the technical assistance of B.Sc. Mario Alimehemeti in the development of the experiments.

Conflicts of Interest: The authors declare no conflict of interest.

References

1. Mbarek, A.; Del Rincon, A.F.; Hammami, A.; Iglesias, M.; Chaari, F.; Viadero, F.; Haddar, M. Comparison of experimental and operational modal analysis on a back to back planetary gear. *Mech. Mach. Theory* **2018**, *124*, 226–247. [CrossRef]
2. Chen, X.; Cheng, G.; Li, H.; Li, Y. Fault identification method for planetary gear based on DT-CWT threshold denoising and LE. *J. Mech. Sci. Technol.* **2017**, *31*, 1035–1047. [CrossRef]
3. Sun, R.; Yang, Z.; Chen, X.; Tian, S.; Xie, Y. Gear fault diagnosis based on the structured sparsity time-frequency analysis. *Mech. Syst. Signal Process.* **2018**, *102*, 346–363. [CrossRef]
4. Zhu, C.; Chen, S.; Liu, H.; Huang, H.; Li, G.; Ma, F. Dynamic analysis of the drive train of a wind turbine based upon the measured load spectrum. *J. Mech. Sci. Technol.* **2014**, *28*, 2033–2040. [CrossRef]
5. Heidari Bafroui, H.; Ohadi, A. Application of wavelet energy and Shannon entropy for feature extraction in gearbox fault detection under varying speed conditions. *Neurocomputing* **2014**, *133*, 437–445. [CrossRef]
6. Chen, X.; Feng, Z. Iterative generalized time–frequency reassignment for planetary gearbox fault diagnosis under nonstationary conditions. *Mech. Syst. Signal Process.* **2016**, *80*, 429–444. [CrossRef]
7. Liu, W.; Shuai, Z.; Guo, Y.; Wang, D. Modal properties of a two-stage planetary gear system with sliding friction and elastic continuum ring gear. *Mech. Mach. Theory* **2019**, *135*, 251–270. [CrossRef]
8. Kong, Y.; Wang, T.; Chu, F. Meshing frequency modulation assisted empirical wavelet transform for fault diagnosis of wind turbine planetary ring gear. *Renew. Energy* **2019**, *132*, 1373–1388. [CrossRef]
9. Li, Y.; Cheng, G.; Liu, C.; Chen, X. Study on planetary gear fault diagnosis based on variational mode decomposition and deep neural networks. *Measurement* **2018**, *130*, 94–104. [CrossRef]
10. Li, Y.; Ding, K.; He, G.; Yang, X. Vibration modulation sidebands mechanisms of equally-spaced planetary gear train with a floating sun gear. *Mech. Syst. Signal Process.* **2019**, *129*, 70–90. [CrossRef]
11. Zghal, B.; Graja, O.; Dziedziech, K.; Chaari, F.; Jablonski, A.; Barszcz, T.; Haddar, M. A new modeling of planetary gear set to predict modulation phenomenon. *Mech. Syst. Signal Process.* **2019**, *127*, 234–261. [CrossRef]
12. Tsai, S.J.; Huang, G.L.; Ye, S.Y. Gear meshing analysis of planetary gear sets with a floating sun gear. *Mech. Mach. Theory* **2015**, *84*, 145–163. [CrossRef]
13. Shen, Z.; Qiao, B.; Yang, L.; Luo, W.; Chen, X. Evaluating the influence of tooth surface wear on TVMS of planetary gear set. *Mech. Mach. Theory* **2019**, *136*, 206–223. [CrossRef]
14. Feki, N.; Karray, M.; Khabou, M.T.; Chaari, F.; Haddar, M. Frequency analysis of a two-stage planetary gearbox using two different methodologies. *C. R. Mec.* **2017**, *345*, 832–843. [CrossRef]
15. Park, J.; Hamadache, M.; Ha, J.M.; Kim, Y.; Na, K.; Youn, B.D. A positive energy residual (PER) based planetary gear fault detection method under variable speed conditions. *Mech. Syst. Signal Process.* **2019**, *117*, 347–360. [CrossRef]
16. Park, J.; Kim, Y.; Na, K.; Youn, B.D. Variance of energy residual (VER): An efficient method for planetary gear fault detection under variable-speed conditions. *J. Sound Vib.* **2019**, *453*, 253–267. [CrossRef]
17. Martins, R.C.; Fernandes, C.M.C.G.; Seabra, J.H.O. Evaluation of bearing, gears and gearboxes performance with different wind turbine gear oils. *Friction* **2015**, *3*, 275–286. [CrossRef]
18. Gani, B.; Ceccarelli, M.; Carbone, G. Design and numerical characterization of a new planetary transmission. *Int. J. Innov. Technol. Res.* **2014**, *2*, 735–739.

Article

A New Stiffness Performance Index: Volumetric Isotropy Index

İbrahimcan Görgülü * and Mehmet İsmet Can Dede

Department of Mechanical Engineering, Izmir Institute of Technology, Urla, İzmir 35430, Turkey; candede@iyte.edu.tr
* Correspondence: ibrahimcangorgulu@iyte.edu.tr

Received: 8 April 2019; Accepted: 17 June 2019; Published: 18 June 2019

Abstract: A new index for a precise calculation of a manipulator's stiffness isotropy is introduced. The proposed index is compared with the conventionally used stiffness isotropy index by making use of the investigation on R-CUBE manipulator. The proposed index is shown to produce relatively more precise results from which a higher number of isotropic poses are detected.

Keywords: stiffness modelling; performance indices; condition number; volumetric isotropy index; parallel manipulator

---

1. Introduction

In industrial robotics, stiffness property has great importance especially when they are used in manufacturing processes. Production quality during milling and drilling operations depends on the stiffness performance of the robot. Thus, many academic studies are devoted to the stiffness analysis and evaluation of industrial robots [1–7].

One reason for the stiffness performance evaluation of a manipulator is to determine the capabilities of manipulators [4,8–11]. There are several evaluation approaches of stiffness properties. One is commonly used in industrial robots that consider the absolute end-effector deflection. In this approach, a specific position or a trajectory is given to the robot, and its positioning error is observed under external wrenches [8,12,13]. The procedures of this kind of evaluation are given in ANSI and ISO standards [4,14]. These methods are suitable to have a general idea on stiffness performance, but they may not distinguish whether forces or moments cause the elastic translational or rotational displacements. Depending on the performance output, a suitable manipulator for the desired task can be selected.

Another reason is to optimally design the manipulator parameters by considering the stiffness performance [5–7,13,15–17]. The optimal design process of a manipulator requires objective functions that include the stiffness performance indices. These indices are mainly used to achieve high rigidity while preserving low inertia and high kinematic performance.

In parallel robots, stiffness modulation can be achieved by introducing redundancy in actuation [18,19] and/or kinematics [20–22]. Stiffness characteristics of the manipulator are regulated by applying suitable redundancy resolution algorithms in control [19–21,23,24]. A common way of redundancy resolution is via null-space control [25] in which a performance index is designed. This performance index is defined as a stiffness performance index when it is required to modulate the stiffness of the manipulator [18–24]. A key aspect of this performance index is that it should be computationally efficient.

Optimal design and control processes require a mathematical stiffness model (stiffness matrix) of the manipulator to use the stiffness performance indices. These indices adopt common matrix operations such as the determinant, norm, or singular value decomposition (SVD) in defining the performance indices [3–6,8–11,13,14,16,17,26–35]. Each of them investigates a different property of the stiffness matrix. For instance, the determinant is accepted as a scalar indication of stiffness magnitude [3,11,14,16,26,28]. SVD computes the eigenvectors and eigenvalues that show the most/least stiff directions and their magnitudes [3,4,6,8,9,11,13,14,17,26–28,30]. Once the stiffness model is known these approaches are quite handy and they provide detailed stiffness information of the manipulator. However, their general problems are that they are calculated locally.

Isotropy is having the same stiffness performance distribution in all directions. In the literature, eigenvalues of the stiffness matrix are used for relative stiffness resolution evaluation. The evaluation is carried out by calculating the ratio of the maximum to the minimum eigenvalue. The index is named as stiffness condition number, stiffness dexterity, or stiffness isotropy index [10,13,14]. It obtains a directional stiffness resolution compared to the maximum eigenvalue. However, only focusing on maximum and minimum eigenvalues hides the effect of intermediate eigenvalues on the performance. For instance, there exist four possible cases for the distribution of eigenvalues of a 3×3 stiffness matrix.The first case is the one that has all different eigenvalues. The second case has two equal minimum eigenvalues. The third case has two equal maximum eigenvalues. The fourth case has all equal eigenvalues. For the first three cases, if the minimum and maximum eigenvalues are the same, then, the stiffness isotropy index computation will output the same value. However, it is clear that the isotropy of first, second, and third cases should be different from each other. This problem can only be resolved by taking into account the intermediate eigenvalues in stiffness isotropy computation.

In this study, special attention is directed towards this intermediate eigenvalue problem. In this regard, a new performance index is proposed. This index composes a volumetric isotropy index by considering intermediate eigenvalues. As a case study, an R-CUBE [36] parallel manipulator is considered. A comparison is conducted between the proposed volumetric isotropy index and existing isotropy index. R-CUBE is selected for this analysis since it has relatively trivial kinematics to be in conveniently used in the formulation of its stiffness model [36].

The remaining sections of this paper are organized as follows; in Section 2 stiffness model of the R-CUBE is obtained, and notations are given. Then, in Section 3, currently used performance indices are discussed, and a new performance index is introduced. In Section 4, results of conventional isotropy index and the new index computations are presented. Finally, conclusions are stated in Section 5.

2. Stiffness Model of R-CUBE

The R-CUBE manipulator is introduced by [36]. The manipulator composed of revolute joints, only. In total, 3 serial chains exist, and each one of them controls their respective translational DoF in Cartesian space. In Figure 1 the kinematic model is illustrated.

In Figure 1a, first refecence frames of the each serial chain are located on $\vec{u}_k^{(0)}$ orthogonal axes along kth axis for $k = 1,2,3$. ij stands for the jth frame in ith serial chain as shown in Figure 1b. $\vec{u}_k^{(0)} \parallel \vec{u}_k^{(p)}$ due the kinematic constraints. p is mobile platform frame. Besides, $\vec{u}_3^{(15)}, \vec{u}_3^{(25)}, \vec{u}_3^{(35)}$ are always aligned with $\vec{u}_1^{(35)}, \vec{u}_1^{(15)}, \vec{u}_1^{(25)}$ vectors, respectively. The forward kinematics is given as:

$$r_i = S + l_1 \sin \varphi_{i1} \ \text{ for } i = 1,2,3 \ \text{ and } \ \bar{r} = \begin{bmatrix} r_1 & r_2 & r_3 \end{bmatrix}^T \tag{1}$$

where S is a constant distance from 0th frame to $\vec{u}_3^{(i0)}$ frame. $\bar{r}$ denotes position vector with respect to the origin.

Commonly used stiffness (elasto-static) modeling methods are classified as Finite Element Method (FEM) [37,38], Matrix Structural Method (MSM) [39,40], and Virtual Joint Method (VJM) [28,41–45]. FEM exhibits the highest accuracy in exchange of computation cost due to its numerical approach. Besides, a meshing operation must be conducted for each pose of the manipulator. VJM and MSM are faster in this evaluation since they construct semi-analytical or analytical models. Thus, stiffness performance metrics are used via VJM and MSM for evaluation of a manipulator in any of its pose. In this study, VJM approach is adopted.

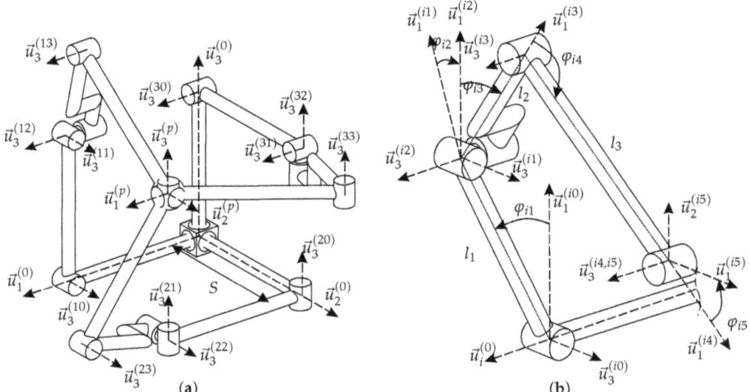

Figure 1. Kinematic sketches of the R-CUBE mechanism: (**a**) The manipulator, (**b**) variables of *i*th serial chain where *i* is one of the serial chains.

Stiffness model of the R-CUBE manipulator is computed by combining the stiffness models of each serial chain. Compliance model of *i*th serial chain and its connection to base and mobile platforms are illustrated in Figure 2. Passive and active joints have 1 degree-of-freedom (DoF) while virtual joints have 6 DoF (3 translations + 3 rotations). A virtual joint is defined as:

$$H_v(\bar{\theta}_{ij}) = T_1(\theta_{ij}^1) T_2(\theta_{ij}^2) T_3(\theta_{ij}^3) R_1(\theta_{ij}^4) R_2(\theta_{ij}^5) R_3(\theta_{ij}^6) \tag{2}$$

where H_v denotes the homogeneous transformation matrix (HTM). T_k and R_k are pure translational HTM along and pure rotational HTM about $\vec{u}_k$th axis for $k = 1, 2, 3$. $\bar{\theta}_{ij}$ is a vector that contains virtual joint variables. Superscripts of θ_{ij} represents the element number.

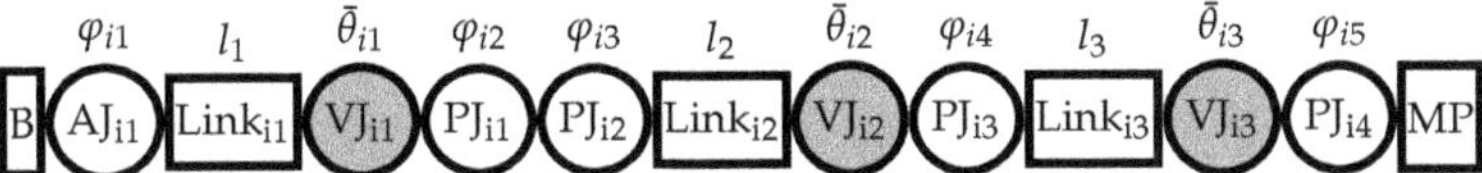

Figure 2. Compliant kinematics of the manipulator where AJ is active joint, PJ is passive joint, VJ is virtual joint, MP is mobile Platform, and B is base.

The compliant kinematic model is computed as follows:

$$
\begin{aligned}
H^{(i0,K_{i1})} &= R_3(\varphi_{i1}) T_{1u}(l_1) H_v(\bar{\theta}_{i1}) \\
H^{(K_{i1},K_{i2})} &= R_3(\varphi_{i2}) R_1(-\pi/2) R_3(\varphi_{i3}) T_1(l_2) H_v(\bar{\theta}_{i2}) \\
H^{(K_{i2},K_{i3})} &= R_3(\varphi_{i4}) T_1(l_3) H_v(\bar{\theta}_{i3}) \\
H^{(K_{i3},i5)} &= R_3(\varphi_{i5}) \\
H_{Ki} &= H^{(0,i0)} H^{(i0,K_{i1})} H^{(K_{i1},K_{i2})} H^{(K_{i2},K_{i3})} H^{(K_{i3},i5)} H^{(i5,p)} \\
H_{Ki} &= \begin{bmatrix} R_{Ki} & \vec{r}_{Ki} \\ \bar{0}^T & 1 \end{bmatrix}
\end{aligned}
\tag{3}
$$

where H_{Ki} is the compliant transformation matrix of the *i*th serial kinematic chain of R-CUBE, and $H_{K1} = H_{K2} = H_{K3}$ by the assumption of a rigid mobile platform.

Active joints are assumed to be rigidly locked to exclude actuation stiffness in the structural stiffness calculations. Then, virtual and passive joint variables are arranged in a matrix form as follows:

$$\bar{q}_{pi} = \begin{bmatrix} \varphi_{i2} & \varphi_{i3} & \varphi_{i4} & \varphi_{i5} \end{bmatrix}^T_{4\times 1}, \quad \bar{\theta}_i = \begin{bmatrix} \bar{\theta}_{i1}^T & \bar{\theta}_{i2}^T & \bar{\theta}_{i3}^T \end{bmatrix}^T_{18\times 1}, \quad \bar{Q}_i = \begin{bmatrix} \bar{\theta}_i^T & \bar{q}_{pi}^T \end{bmatrix}^T_{22\times 1} \tag{4}$$

where $\bar{q}_{pi}$ denotes passive joint variables, $\bar{Q}_i$ is generalized coordinates of ith serial chain. $\bar{Q}_i$ is used to obtain Jacobian matrices as follows:

$$\frac{\partial H_{Ki}}{\partial Q_{ik}} = \begin{bmatrix} \frac{\partial R_{Ki}}{\partial Q_{ik}} & \frac{\partial \bar{r}_{Ki}}{\partial Q_{ik}} \\ \bar{0}^T & 1 \end{bmatrix} \text{ for } k = 1, 2, ..., 22 \tag{5}$$

where subscript k denotes the kth variable of $\bar{Q}_i$.

The Jacobian matrix J_{Ki} for passive and virtual joints is given as follows:

$$J_{Ki} = \begin{bmatrix} \bar{J}_{Ki1} & \bar{J}_{Ki2} & \cdots & \bar{J}_{Ki22} \end{bmatrix}_{6\times 22} \tag{6}$$

where $\bar{J}_{Kik}$ for $k = 1, 2, ..., 22$ denotes the Jacobian column matrices that are related with k th variable of Q_i.

J_{Ki} can be divided into sub-matrices as $J_{\theta i}$ and J_{pi}. They denote Jacobian matrices of virtual and passive joints. Sub-matrices are presented as follows:

$$J_{\theta i} = \begin{bmatrix} \bar{J}_{Ki1} & \bar{J}_{Ki2} & \cdots & \bar{J}_{Ki18} \end{bmatrix}_{6\times 18}$$
$$J_{pi} = \begin{bmatrix} \bar{J}_{Ki19} & \bar{J}_{Ki20} & \bar{J}_{Ki21} & \bar{J}_{Ki22} \end{bmatrix}_{6\times 4} \tag{7}$$
$$J_{Ki} = \begin{bmatrix} J_{\theta i} & J_{pi} \end{bmatrix}_{6\times 22}$$

Note that, except in a kinematically singular configuration, $J_{\theta i}$ is always full-rank since the virtual joints have decoupled DoF. Similarly, J_{pi} gets rank deficient in kinematic singularities.

Jacobian matrices relate small deflections of joints to task space variables:

$$\bar{X}_i = \bar{f}(\bar{Q}_i), \quad \Rightarrow \Delta \bar{X}_i = J_{Ki}\Delta\bar{Q}_i$$
$$\Delta \bar{X}_i = J_{\theta i}\Delta\bar{\theta}_i + J_{pi}\Delta\bar{q}_{pi} \tag{8}$$

where $\Delta \bar{X}_i$ is 6×1 column matrix of translational and rotational compliant deflections in task space that are calculated for the ith serial chain. Δ operator denotes the change between initial and final states. In the initial state, there is no applied wrench. In the final state, an external wrench is applied to the manipulator.

The relation between the external force/torque applied at the end-effector and joint space force/torque is provided via a property of Jacobian matrix that is given as follows:

$$\bar{F}_{Ki} = J_{Ki}^T\bar{F}_{ext} \tag{9}$$

where $[\bar{F}_{Ki}]_{22\times 1}$ is the joint space force/torque vector. $[\bar{F}_{ext}]_{6\times 1}$ is the external wrench. $\bar{F}_{Ki}$ is divided into sub-components and the force/torque vector on each joint are found as follows:

$$\bar{F}_{Ki} = \begin{bmatrix} \bar{F}_{\theta i}^T & \bar{F}_{pi}^T \end{bmatrix}^T$$
$$\begin{bmatrix} \bar{F}_{\theta i}^T & \bar{F}_{pi}^T \end{bmatrix}^T = \begin{bmatrix} J_{\theta i} & 0 \end{bmatrix}^T \bar{F}_{ext} + \begin{bmatrix} 0 & J_{pi} \end{bmatrix}^T \bar{F}_{ext} \tag{10}$$

where $\bar{F}_{\theta i}$ and $\bar{F}_{pi}$ are force/torque vectors of virtual and passive joints.

$\bar{F}_{Ki}$ is a function of the stiffness matrix and deflections that are defined in joints space.

$$\bar{F}_{Ki} = \text{diag}(K_{\theta i}, K_{pi})\Delta\bar{Q}_i \quad \text{and} \quad K_{\theta i} = \text{diag}(K_{\theta i1}, K_{\theta i2}, K_{\theta i3}) \tag{11}$$

where $K_{\theta ik}$ denotes stiffness matrix of kth link as expressed in [46] for a beam. Parameters of this stiffness matrix depends on the link geometry and material property. $K_{\theta i}$ denotes the stiffness matrix of ith serial chain. K_{pi} denotes stiffness of passive joints. A relation is expressed in Cartesian space as follows:

$$J_{Ki}^T \bar{F}_{ext} = \text{diag}(K_{\theta i}, K_{pi})\Delta\bar{Q}_i \quad \text{and} \quad J_{Ki}^{-1}\Delta\bar{X}_i = \Delta\bar{Q}_i$$
$$\Rightarrow \bar{F}_{ext} = (J_{\theta i}^{-T} K_{\theta i} J_{\theta i}^{-1} + J_{pi}^{-T} K_{pi} J_{pi}^{-1})\Delta\bar{X}_i \tag{12}$$

Passive joints do not generate reaction torques about their rotation axes. Hence, $K_{pi} = 0$. Thus, $\bar{F}_{Ki}$ contains only force/torque of virtual joints. Hence:

$$\bar{F}_{ext} = (J_{\theta i}^{-T} K_{\theta i} J_{\theta i}^{-1})\Delta\bar{X}_i$$
$$J_{pi}^T \bar{F}_{ext} = \bar{0} \tag{13}$$

The effect of passive joints is included in the stiffness model by inverting the following homogeneous relation matrix. This matrix is always invertible if $\det(J_{\theta i}^T J_{\theta i}) \neq 0$.

$$\begin{bmatrix} (J_{\theta i} K_{\theta i}^{-1} J_{\theta i}^T) & J_{pi} \\ J_{pi}^T & 0 \end{bmatrix}^{-1} = \begin{bmatrix} [K_{Ci}]_{6\times 6} & \sim \\ \hline \sim & \sim \end{bmatrix} \tag{14}$$

where K_{Ci} denotes stiffness matrix of ith in Cartesian space. Note that, K_{Ci} is rank deficient due to the passive joints. Cartesian stiffness matrix of the manipulator, K_C, is computed as $K_C = \sum_{i=1}^{3} K_{Ci}$. If external wrench is assumed to be $\bar{F}_{ext} = \bar{0}$, K_C takes the following form.

$$K_C = \begin{bmatrix} K_{C_1}^{(11)} & 0 & 0 & 0 & K_{C_1}^{(15)} & K_{C_1}^{(16)} \\ 0 & K_{C_2}^{(22)} & 0 & K_{C_2}^{(24)} & 0 & K_{C_2}^{(26)} \\ 0 & 0 & K_{C_3}^{(33)} & K_{C_3}^{(34)} & K_{C_3}^{(35)} & 0 \\ 0 & K_{C_2}^{(24)} & K_{C_3}^{(34)} & K_{C_2}^{(44)} + K_{C_3}^{(44)} & K_{C_3}^{(45)} & K_{C_2}^{(46)} \\ K_{C_1}^{(15)} & 0 & K_{C_3}^{(35)} & K_{C_3}^{(45)} & K_{C_1}^{(55)} + K_{C_3}^{(55)} & K_{C_1}^{(56)} \\ K_{C_1}^{(16)} & K_{C_2}^{(26)} & 0 & K_{C_2}^{(46)} & K_{C_1}^{(56)} & K_{C_1}^{(66)} + K_{C_2}^{(66)} \end{bmatrix} \tag{15}$$

For small deflections and loads, this matrix can be used without causing high errors. This matrix must be re-computed if $|\bar{F}_{ext}| >> 0$. K_C is divided into 3×3 sub-matrices.

$$K_C = \begin{bmatrix} K_A & K_B \\ K_B^T & K_D \end{bmatrix} \tag{16}$$

where K_A, K_B, K_B^T, and K_D have the units of N/m, N/rad, N/rad, and Nm, respectively. The kinematic dimensions and material properties of the links are given in [47]. This study makes use of these parameters for stiffness evaluation of R-CUBE manipulator.

3. Stiffness Performance Indices

In this section, a review of the existing performance indices in literature is described, and a new performance index for isotropy evaluation of the stiffness matrix is proposed. Besides, usage of the indices for the R-CUBE manipulator is given in a table.

3.1. Performance Indices in the Literature

Stiffness performance indices are computed by obtaining the eigenvalues and eigenvectors. SVD operation may be used to obtain these properties as follows:

$$K = QDU \tag{17}$$

where K is an $n \times n$ stiffness matrix defined in Cartesian space for $n = 1, 2, ..., 6$. Q and U are orthogonal matrices whose columns are the eigenvectors of K. Since K is a symmetric matrix, $Q = U^T$. D is a diagonal matrix whose elements denote the positive eigenvalues. These matrices are shown as:

$$Q = \begin{bmatrix} \bar{e}_1 & \bar{e}_2 & ... & \bar{e}_6 \end{bmatrix} \tag{18}$$

$$D = diag(\lambda_1, \lambda_2, ..., \lambda_6) \tag{19}$$

$$\lambda_1 \leq \lambda_2 \leq ... \leq \lambda_6 \tag{20}$$

where $\bar{e}_i$ and λ_i for $i = 1, 2, ..., 6$ denote the eigenvectors and eigenvalues, respectively. Regardless of the kinematic DoF of the manipulator, stiffness matrix is always a 6 × 6 matrix because compliant displacement may occur in any direction in 6 DoF space. Direct evaluation of a 6 × 6 stiffness matrix causes unit inconsistency in the results. A simple method is to use the normalized stiffness matrix [48]. The proposed method uses pre- and post-multiplication of the matrix with diagonal scaling matrices. Another method is introduced by Angeles [49]. He used the natural length to obtain normalized and unity matrix. Plücker Coordinates is preferred by Khan and Angeles to use dimensionally homogeneous space [50]. Thus, the dimensionally homogeneous matrix is obtained. A separate evaluation of rotation sensitive and translation sensitive matrices are introduced by Cardou et al. [51]. Hence, it is also possible to use interested sub-matrices of 6 × 6 stiffness matrix. In this way, a certain aspect of the stiffness model can be placed in focus. In our case, since we focus on the translational compliant displacements of the mobile platform, we used the top left corner of the stiffness matrix as indicated in Equation (16). Hence, in this sub-matrix, unit consistency is achieved. Nevertheless, this produces a lower dimensional stiffness matrix.

The graphical illustration of eigenvectors and eigenvalues generates an n-dimensional surface. This surface becomes a line for $n = 1$, an ellipse for $n = 2$, an ellipsoid for $n = 3$, and a hyper-ellipsoid for $n = 4, 5, 6$. The radii and axes of ellipsoids are defined by eigenvectors and eigenvalues as shown in Figure 3 for $n = 3$ [26].

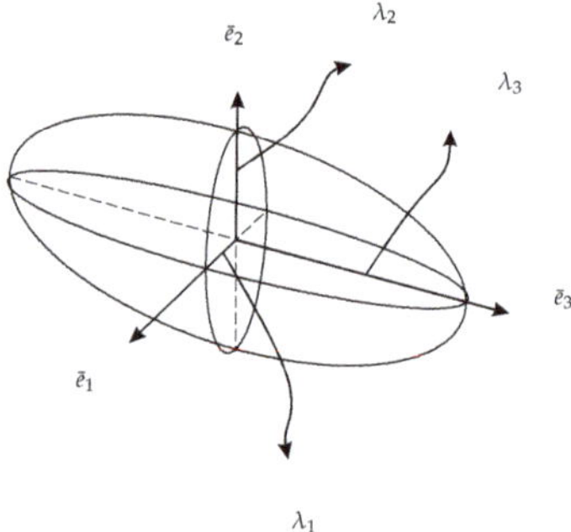

Figure 3. Illustration of eigenvalues (λ_i for $i = 1, 2, 3$) and eigenvectors ($\bar{e}_i$ for $i = 1, 2, 3$) as an ellipsoid for a 3 × 3 stiffness matrix.

Eigenvalues indicate the magnitude of the stiffness along their respective eigenvectors. Hence, the volume of this ellipsoid indicates the stiffness capacity. This volume is proportional to the determinant of the stiffness matrix [3,11,14,16,26–28,32]. Thus, stiffness value is commonly computed as:

$$\gamma_V = \det(\mathbf{K}) = \prod_{i=1}^{n} \lambda_i \tag{21}$$

where γ_V denotes an average stiffness magnitude, and λ_i is the ith eigenvalue. γ_V equals to zero when inspected pose of the manipulator has at least one free motion direction. This case corresponds to a stiffness singularity. Thus, such a singular pose can be determined by investigating γ_V.

The Euclidean norm $\|.\|_E$, which is also named as 2-norm $\|.\|_2$, computes the square root of the largest positive eigenvalue (or singular value) of the square of a matrix, $\mathbf{K}\mathbf{K}^T$ shown as follows:

$$\|\mathbf{K}\|_E = \max(\sqrt{\tilde{\lambda}_i}) \tag{22}$$

where $\tilde{\lambda}_i$ denotes the ith eigenvalue of matrix $\mathbf{K}\mathbf{K}^T$. This norm exhibits the maximum eigenvalue of $\mathbf{K}$ which is also the largest radius of the ellipsoid. The direction of this eigenvalue is the direction of the displacement which exhibits the highest stiffness. Euclidean norm of $\mathbf{K}^{-1}$ reveals the minimum eigenvalue of $\mathbf{K}$, which indicates the most compliant direction [14,17,27,29,30,33]. In optimization problems, it is common to only focus on minimum eigenvalue in the workspace in order to maximize it.

Other norms are 1 norm $\|.\|_1$, infinity norm $\|.\|_\infty$ (also named as Chebyshev norm $\|.\|_C$), and Frobenius norm $\|.\|_F$. 1 norm computes the maximum of the summation of absolute values of column elements of a matrix. Infinity norm makes this computation for rows. Since the stiffness matrix is symmetric, both norms result in the same values. These norms denote a combined total resistive force and moment against a unit displacement in Cartesian space. Frobenius norm, on the other hand, has more meaning in terms of average stiffness. It focuses on diagonal elements of $\mathbf{K}\mathbf{K}^T$. It is computed as follows:

$$\|\mathbf{K}\|_F = \left(\sum_{i=1}^{n} \sum_{j=1}^{n} (K_{ij})^2 \right)^{1/2} = \sqrt{tr(\mathbf{K}\mathbf{K}^T)} = \sqrt{\lambda_1^2 + \lambda_2^2 + ... + \lambda_n^2} \tag{23}$$

where tr is the trace operator and subscript ij denotes the ijth element of the matrix. All norms result in higher values in stiff poses and lower values in compliant poses. Hence, they show similar relative distribution throughout the workspace. Therefore, an evaluation of one of the norms is sufficient to have an idea of stiffness distribution depending on manipulator pose.

Stiffness condition number, stiffness dexterity, or stiffness isotropy index is computed as the ratio of the maximum eigenvalue to the minimum eigenvalue [3,6,10,14,17,26,28,31]. This ratio reveals stiffness value distribution among eigenvectors. The minimum ratio of 1 indicates equal stiffness distribution. The calculation of this index is given as:

$$\gamma_C = \frac{\lambda_n}{\lambda_1} = \|\mathbf{K}\|_E \|\mathbf{K}^{-1}\|_E \tag{24}$$

where γ_C is stiffness condition number.

Another stiffness performance index is the uniformity index which compares the maximum and minimum values of γ_C. This index is formulated as follows:

$$\gamma_{CU} = \frac{max(\gamma_C)}{min(\gamma_C)} \tag{25}$$

$$\gamma_{CU} \geq 1 \tag{26}$$

where γ_{CU} is the uniformity index. Notice that, the minimum value of γ_C is bounded by 1 and so $\gamma_{CU} \geq 1$. Here, $\gamma_{CU} = 1$ can only be achieved when the maximum and minimum values of γ_C are equal to each other. The only condition that satisfies this equality is when all γ_C in all poses are equal to each other.

The determination of both stiff and isotropic poses is a problem. A solution is proposed in [15] that evaluates the stiffness magnitude and isotropy, simultaneously. The formulation of this index is given below.

$$\gamma_G = \frac{\lambda_n^2 \lambda_1^2}{\lambda_n + \lambda_1} \tag{27}$$

γ_G value increases as the manipulator approaches a stiff and isotropic pose. This index is more suitable for 2×2 stiffness matrices. For higher dimensions, this approach is not appropriate because the problem turns into a volumetric problem while γ_G solves a surface equation.

Energy index computation depends on whether a constant payload is applied or a constant compliant displacement is given to a manipulator. For a constant payload, stiff poses of the manipulator result in smaller deflections. Since the energy has a quadratic relationship with elastic displacements, stiff manipulators store less energy for the same payload. Therefore, the highest stored energy is observed when the elastic displacement is along the eigenvector direction that has the minimum eigenvalue for a payload. On the other hand, when a constant compliant displacement is given to manipulator, obviously stiff poses store more energy. It means the highest energy for a constant displacement is observed when this displacement is given along the eigenvector that has the highest eigenvalue. For a unit displacement, stored energy equals to the half of the maximum eigenvalue. The index formulation is given as:

$$\gamma_\epsilon = \frac{1}{2}\Delta \bar{X}^T K \Delta \bar{X} \ \Rightarrow \ \gamma_\epsilon = \frac{1}{2}\bar{e}_n^T K \bar{e}_n \ \Rightarrow \ \gamma_\epsilon = \frac{1}{2}\lambda_n \ \Rightarrow \ \gamma_\epsilon = \frac{1}{2}\|K\|_E \tag{28}$$

where γ_ϵ is the energy index [5,7,11,52].

Unfortunately, all the above indices are computed numerically. In addition, they are local indices (except γ_{Cu}). Hence, they are pose dependent, and they only indicate local performances. Thus, a globalization process is necessary for these indices [35]. This globalization can be achieved for the abovementioned indices to obtain an average value of performance of the whole workspace as follows:

$$\bar{\mu} = \frac{\int \mu dv}{\int dv} \tag{29}$$

$$\sigma = \sqrt{\frac{\int (\mu - \bar{\mu})^2 dv}{\int dv}} \tag{30}$$

where μ denotes any of the performance indices. $\bar{\mu}$ is an average value, and σ is the standard deviation. It is desired to have a low deviation to have a uniform stiffness performance distribution.

3.2. Proposed Performance Index

As can be seen in Section 3.1, many of the indices mostly rely on the maximum and minimum eigenvalues for performance evaluation. However, this causes a lack in proper stiffness isotropy evaluation for $n \times n$ dimensional stiffness matrix where $n > 2$. In this respect, a revision is required for stiffness isotropy index.

While stiffness condition number for a 2×2 stiffness matrix regarded as the ratios of radii of a stiffness ellipse, the mathematical meaning behind is the comparison of the areas of a circle and an ellipse. Area of this circle indicates an ideal (desired) performance value which has high isotropy and rigidity. Hence, the square of the maximum eigenvalue is an indication of the area of the circle. Area of the ellipse is the multiplication of all eigenvalues (determinant). The ratio between these areas gives the stiffness condition number. For a 2×2 stiffness matrix, this index is formulated as follows:

$$\gamma_C = \frac{\lambda_2^2}{\lambda_1 \lambda_2} = \frac{\lambda_2}{\lambda_1} \tag{31}$$

$$\gamma_C \geq 1 \tag{32}$$

The current γ_C is a 2 DoF evaluation approach and has a lack of performance in the evaluation of 3 or more DoF problems. Therefore, γ_C cannot distinguish the performances of poses whose eigenvalues $\lambda_1 = \lambda_2 < \lambda_3$, $\lambda_1 < \lambda_2 = \lambda_3$, or $\lambda_1 < \lambda_2 < \lambda_3$ for 3×3 matrix. Poses that have closer eigenvalues to maximum one are more isotropic. Hence, γ_C must be revised for higher DoF evaluation. Higher dimensional isotropy index is the comparison of volumes of an ideal n-dimensional sphere and an n-dimensional ellipsoid. In this regard, the extension of the condition number for $n \times n$ stiffness matrix is a volumetric condition number or volumetric isotropy index that is formulated as follows:

$$\gamma_I = \frac{\|K\|_E^n}{\det(K)} \tag{33}$$

$$\gamma_I \geq 1 \tag{34}$$

where γ_I volumetric stiffness isotropy index and n is the dimension of the stiffness matrix, nth power of $\|K\|_E$ indicates the volume of the ideal sphere while $\det(K)$ represented the volume of the ellipsoid. $\gamma_I = 1$ indicates the isotropic poses in terms of stiffness. Since this index also considers the intermediate eigenvalue contribution, the ambiguity of stiffness performance between the poses that have same γ_C can be accurately distinguished. Hence, it is expected to have a lower standard deviation of σ_I compared to σ_C. The standard deviation in such a comparison can be regarded as the preciseness of the performance index.

A volumetric uniformity index γ_{IU} may also be defined as an extension to γ_I. The uniformity index is defined as follows:

$$\gamma_{IU} = \frac{max(\gamma_I)}{min(\gamma_I)} \tag{35}$$

$$\gamma_{IU} \geq 1 \tag{36}$$

Since the preciseness of the volumetric stiffness isotropy index is expected to be relatively better, the volumetric uniformity index is expected to be more accurate compared to γ_{CU}.

3.3. Construction of Performance Indices for R-CUBE

Previously mentioned indices are utilized for the stiffness evaluation R-CUBE manipulator, and they are tabulated in Table 1. Since the mobile platform of the manipulator has only translational DoF, translational compliant displacements are our primary interest. Hence, K_A sub-matrix is in the focal point of this study.

Table 1. Utilized stiffness performance indices.

Index	Matrix	Computed Function	$\bar{\mu}$	σ	Unit
γ_V	K_A	$\det(K_A)$	$\bar{\mu}_V$	σ_V	$(N/m)^3$
$\|\cdot\|_E$	K_A	$\|K_A\|_E$	$\bar{\mu}_E$	σ_E	(N/m)
$\|\cdot\|_1$	K_A	$\|K_A\|_1$	$\bar{\mu}_1$	σ_1	(N/m)
$\|\cdot\|_\infty$	K_A	$\|K_A\|_\infty$	$\bar{\mu}_\infty$	σ_∞	(N/m)
$\|\cdot\|_F$	K_A	$\|K_A\|_F$	$\bar{\mu}_F$	σ_F	(N/m)
γ_ϵ	K_A	$\frac{1}{2}\|K_A\|_E$	$\bar{\mu}_\epsilon$	σ_ϵ	(Joule)
γ_I	K_A	$\|K_A\|_E^3/\det(K_A)$	$\bar{\mu}_I$	σ_I	-
γ_C	K_A	$\|K_A\|_E/\|K_A^{-1}\|_E$	$\bar{\mu}_C$	σ_C	-
γ_{IU}	K_A	$max(\gamma_I)/min(\gamma_I)$	-	-	-
γ_{CU}	K_A	$max(\gamma_C)/min(\gamma_C)$	-	-	-

4. Results and Discussion

In this section, only the results on isotropy index calculations are given to present a comparison between the conventional and proposed isotropy index. γ_I, and γ_C values are computed throughout the workspace, and their results are compared with each other. These computations are carried out by using K_A sub-matrix. A normal distribution plot is presented for both indices to illustrate the relative preciseness of the proposed index γ_I. Isotropy indices are computed for each discrete pose of the workspace. Then, the computed indices are illustrated via color mapping.

The maximum and minimum values, average values, standard deviations of isotropy indices are presented in Table 2. Having the minimum value of 1 for γ_C and γ_I indices shows that there at least one isotropic pose. By definition of both indices, the isotropy is observed in the same poses. Even though both indices compute the same stiffness matrix, their maximum values differ from each other as expected. The effect of intermediate eigenvalue causes this difference. These maximum values indicate the worst isotropy performance. The number of worst isotropic poses with respect to γ_C is higher compared to γ_I because γ_C does not consider the intermediate eigenvalues. However, the worst isotropic poses concerning γ_I are also the worst isotropic poses with respect to γ_C, but vice-versa is not necessarily valid for all least isotropic poses.

The values of γ_C and γ_I are normalized as γ_C^* and γ_I^* to carry out a fair comparison among them. Accordingly, $\bar{\mu}_I^*$, $\bar{\mu}_C^*$, σ_I^*, and σ_C^* are obtained in a normalized space. γ_C and γ_I are normalized such that the worst performance is denoted by 1 and the highest isotropy is given by 0.

Table 2. Results of isotropy indices.

	Max	Min	$\bar{\mu}$	σ
γ_I	5.45	1	$\bar{\mu}_I = 1.80$	$\sigma_I = 0.68$
γ_C	2.37	1	$\bar{\mu}_C = 1.42$	$\sigma_C = 0.26$
γ_{Iu}	5.45	5.45	-	-
γ_{Cu}	2.37	2.37	-	-
γ_I^*	1	0	$\bar{\mu}_I^* = 0.19$	$\sigma_I^* = 0.15$
γ_C^*	1	0	$\bar{\mu}_C^* = 0.31$	$\sigma_C^* = 0.19$

The normal distribution of both indices is compared in a normalized space and shown in Figure 4. Probability density in the vertical axis indicates the number of poses that have the isotropy value denoted in the horizontal axis. Mean values μ_I^*, and μ_C^* indicates the average isotropy of the workspace. γ_I^* results indicate that the workspace is more isotropic with respect to γ_C^* results since $\mu_I^* < \mu_C^*$. In addition, the probability distribution of isotropic poses is higher by using γ_I index. The standard deviation σ_I^* is lower than σ_C^*. Hence, γ_I^* is more sensitive to isotropy changes between poses, and this new index can detect slightest changes.

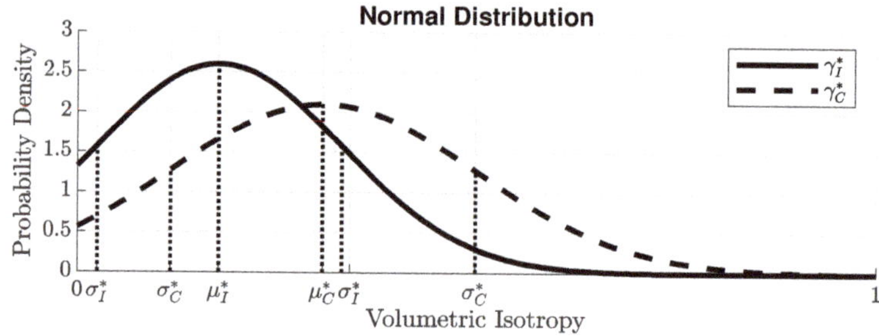

Figure 4. Normal distribution of γ_I and γ_C.

For illustration purposes, 9 planes are selected in the workspace. These planes are parallel to $\vec{u}_1 - \vec{u}_2$, $\vec{u}_2 - \vec{u}_3$, and $\vec{u}_1 - \vec{u}_3$ orthogonal planes. They are divided into 3 groups. In each group, 3 planes are intersect at $[-60, -60, -60]$ mm, $[0, 0, 0]$ mm, $[60, 60, 60]$ mm locations of the workspace. Figures 5a and 6a, show the outermost surfaces of the workspace where the planes are placed at $[60, 60, 60]$ mm. Figures 5b and 6b, show the planes that are coincident at $[0, 0, 0]$ mm. Figures 5c and 6c, show the most inner surfaces of the workapce at $[-60, -60, -60]$ mm. Due to the symmetric topology of the manipulator, computed performance indices have the same value in several poses. Hence, a symmetric distribution is observed in the results of both indices throughout the workspace.

Due to the definition of both indices, lower values (represented with blue zones) indicate higher isotropy. These areas mainly are observed at the innermost corner, the outermost corner, and at the center of the workspace. However, isotropy distribution differs for γ_I^* and γ_C^* for the rest of the workspace. The transition between low isotropy and high isotropy is more smooth in γ_C^* compared to γ_I^*. This smooth transition is a consequence of relatively less precise results obtained with γ_C. Hence, γ_C cannot precisely distinguish the isotropy levels among poses. However, the transition from low to high isotropy in γ_I^* results illustration is sharper and more distinguishable compared to γ_C^*.

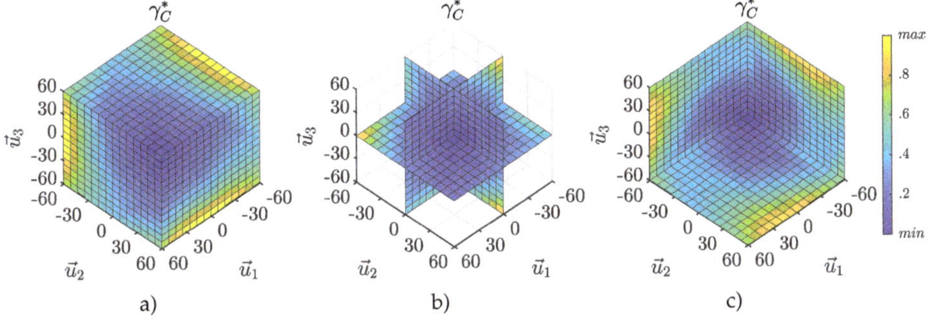

Figure 5. Isotropy distribution: (**a**) Outer surfaces, (**b**) middle surfaces, (**c**) inner surfaces.

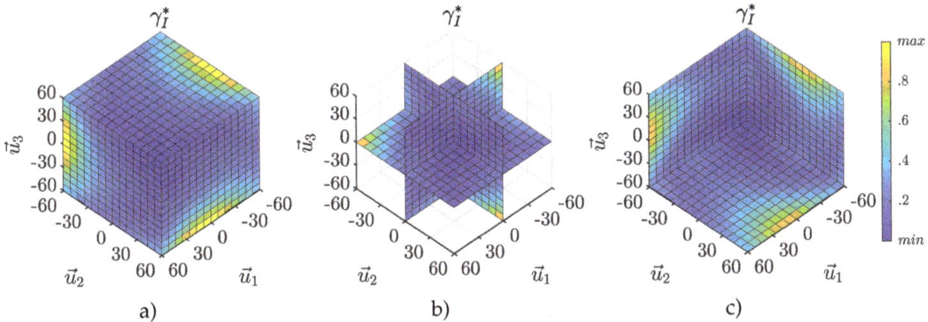

Figure 6. Volumetric isotropy distribution: (**a**) Outer surfaces, (**b**) middle surfaces, (**c**) inner surfaces.

Figures 7 and 8 show the iso-curves in $\vec{u}_1 - \vec{u}_2$ planes for γ_C^*, and γ_I^*, respectively. These planes are positioned along $\vec{u}_3$ and they are located at 60 mm, 0 mm, -60 mm. Figure 8 shows that the covered area by the iso-curves for $\gamma_I^* = 0.1, 0.2$ is higher than the area of $\gamma_C^* = 0.1, 0.2$ in Figure 7. Hence, γ_I captures more isotropic poses. Accordingly, iso-curves for $\gamma_C^* = 0.3, 0.4, ..., 0.8$ cover more area than the area covered by the same value of γ_I^*. This indicates that the manipulator is less isotropic with respect to γ_C. In addition, notice that the distance between the iso-curves are less for γ_I^* in Figure 8 compared to γ_C^* in Figure 7. Therefore, γ_I is more sensitive to isotropy changes.

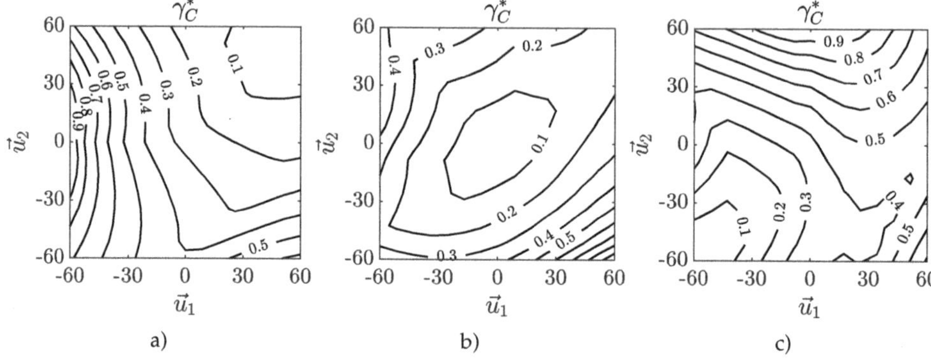

Figure 7. Normalized isotropy iso-curves: (**a**) 60 mm along $\vec{u}_3$, (**b**) 0 mm along $\vec{u}_3$, (**c**) -60 mm along $\vec{u}_3$.

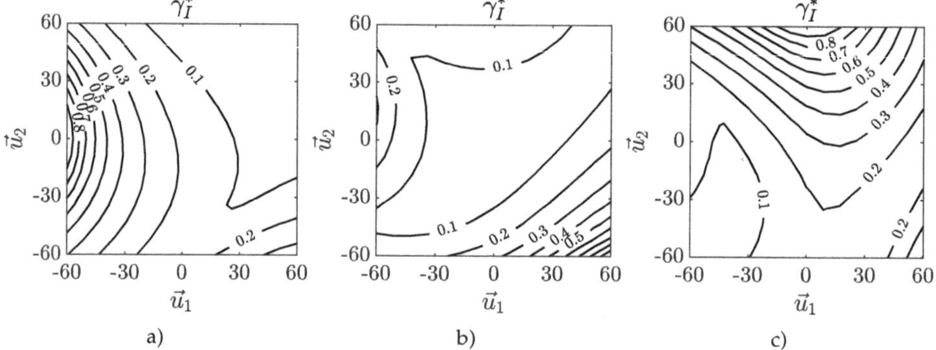

Figure 8. Normalized volumetric isotropy iso-curves: (**a**) 60 mm along $\vec{u}_3$, (**b**) 0 mm along $\vec{u}_3$, (**c**) -60 mm along $\vec{u}_3$.

5. Conclusions

In this study, a stiffness model for the R-CUBE manipulator is derived via VJM. A volumetric isotropy index was proposed and compared with the isotropy index that is used in the literature. The proposed index enables a precise evaluation of n-dimensional stiffness problem. The comparison of indices was achieved in a normalized domain. The isotropy distribution is illustrated in 3-dimensional figures using color mapping and by sketching iso-curves. In addition, the normal distribution is given for both indices. It was observed that the proposed isotropy index is more sensitive than the conventional one. It has a lower standard deviation. Hence, it can distinguish isotropic and non-isotropic poses for spatial stiffness matrices, precisely.

While in this paper VJM is chosen to compute the stiffness model, other stiffness model computation can be employed, and the proposed stiffness index and the other indices can be calculated by using these models as well. As a future study, an optimal design will be conducted for R-CUBE mechanism by using the volumetric isotropy index.

Author Contributions: All the work presented in this paper was carried out by İ.G. under the supervision of M.İ.C.D.

Funding: The study is supported by The Scientific and Technological Research Council of Turkey via grant number 117M405.

Conflicts of Interest: The authors declare no conflict of interest.

References

1. Lian, B.; Sun, T.; Song, Y.; Jin, Y.; Price, M. Stiffness analysis and experiment of a novel 5-DoF parallel kinematic machine considering gravitational effects. *Int. J. Mach. Tools Manuf.* **2015**, *95*, 82–96. [CrossRef]
2. Dong, C.; Liu, H.; Yue, W.; Huang, T. Stiffness modeling and analysis of a novel 5-DOF hybrid robot. *Mech. Mach. Theory* **2018**, *125*, 80–93. [CrossRef]
3. Guo, Y.; Dong, H.; Ke, Y. Stiffness-oriented posture optimization in robotic machining applications. *Robot. Comput.-Integr. Manuf.* **2015**, *35*, 69–76. [CrossRef]
4. Klimchik, A.; Pashkevich, A. Serial vs. quasi-serial manipulators: Comparison analysis of elasto-static behaviors. *Mech. Mach. Theory* **2017**, *107*, 46–70. [CrossRef]
5. Sun, T.; Lian, B. Stiffness and mass optimization of parallel kinematic machine. *Mech. Mach. Theory* **2018**, *120*, 73–88. [CrossRef]
6. Courteille, E.; Deblaise, D.; Maurine, P. Design optimization of a delta-like parallel robot through global stiffness performance evaluation. In Proceedings of the 2009 IEEE/RSJ International Conference on Intelligent Robots and Systems, St. Louis, MO, USA, 11–15 October 2009; pp. 5159–5166.
7. Wang, H.; Zhang, L.; Chen, G.; Huang, S. Parameter optimization of heavy-load parallel manipulator by introducing stiffness distribution evaluation index. *Mech. Mach. Theory* **2017**, *108*, 244–259. [CrossRef]
8. Taghvaeipour, A.; Angeles, J.; Lessard, L. On the elastostatic analysis of mechanical systems. *Mech. Mach. Theory* **2012**, *58*, 202–216. [CrossRef]
9. Arsenault, M. Workspace and stiffness analysis of a three-degree-of-freedom spatial cable-suspended parallel mechanism while considering cable mass. *Mech. Mach. Theory* **2013**, *66*, 1–13. [CrossRef]
10. Gosselin, C.M. Dexterity indices for planar and spatial robotic manipulators. In Proceedings of the IEEE International Conference on Robotics and Automation, Cincinnati, OH, USA, 13–18 May 1990; pp. 650–655.
11. Yan, S.; Ong, S.; Nee, A. Stiffness analysis of parallelogram-type parallel manipulators using a strain energy method. *Robot. Comput.-Integr. Manuf.* **2016**, *37*, 13–22. [CrossRef]
12. Raoofian, A.; Taghvaeipour, A.; Kamali, A. On the stiffness analysis of robotic manipulators and calculation of stiffness indices. *Mech. Mach. Theory* **2018**, *130*, 382–402. [CrossRef]
13. Abdolshah, S.; Zanotto, D.; Rosati, G.; Agrawal, S.K. Optimizing stiffness and dexterity of planar adaptive cable-driven parallel robots. *J. Mech. Robot.* **2017**, *9*, 031004. [CrossRef]
14. Carbone, G.; Ceccarelli, M. Comparison of indices for stiffness performance evaluation. *Front. Mech. Eng. China* **2010**, *5*, 270–278. [CrossRef]
15. Yeo, S.; Yang, G.; Lim, W. Design and analysis of cable-driven manipulators with variable stiffness. *Mech. Mach. Theory* **2013**, *69*, 230–244. [CrossRef]
16. Xu, Q.; Li, Y. GA-based architecture optimization of a 3-PUU parallel manipulator for stiffness performance. In Proceedings of the 2006 6th World Congress on Intelligent Control and Automation, Dalian, China, 21–23 June 2006; Volume 2, pp. 9099–9103.
17. Liu, X.J.; Jin, Z.L.; Gao, F. Optimum design of 3-DOF spherical parallel manipulators with respect to the conditioning and stiffness indices. *Mech. Mach. Theory* **2000**, *35*, 1257–1267. [CrossRef]
18. Lee, S.H.; Lee, J.H.; Yi, B.J.; Kim, S.H.; Kwak, Y.K. Optimization and experimental verification for the antagonistic stiffness in redundantly actuated mechanisms: A five-bar example. *Mechatronics* **2005**, *15*, 213–238. [CrossRef]
19. Pitt, E.B.; Simaan, N.; Barth, E.J. An investigation of stiffness modulation limits in a pneumatically actuated parallel robot with actuation redundancy. In Proceedings of the ASME/BATH 2015 Symposium on Fluid Power and Motion Control, Chicago, IL, USA, 12–14 October 2015; p. V001T01A063.
20. Zhou, X.; Jun, S.k.; Krovi, V. Stiffness modulation exploiting configuration redundancy in mobile cable robots. In Proceedings of the 2014 IEEE International Conference on Robotics and Automation (ICRA), Hong Kong, China, 31 May–7 June 2014; pp. 5934–5939.
21. Alamdari, A.; Haghighi, R.; Krovi, V. Stiffness modulation in an elastic articulated-cable leg-orthosis emulator: Theory and experiment. *IEEE Trans. Robot.* **2018**, *99*, 1–14. [CrossRef]
22. Orekhov, A.L.; Simaan, N. Directional Stiffness Modulation of Parallel Robots with Kinematic Redundancy and Variable Stiffness Joints. *J. Mech. Robot.* **2019**, 1–14. [CrossRef]

23. Ajoudani, A.; Tsagarakis, N.G.; Bicchi, A. On the role of robot configuration in cartesian stiffness control. In Proceedings of the 2015 IEEE International Conference on Robotics and Automation (ICRA), Seattle, WA, USA, 26–30 May 2015; pp. 1010–1016.

24. Ficuciello, F.; Romano, A.; Villani, L.; Siciliano, B. Cartesian impedance control of redundant manipulators for human-robot co-manipulation. In Proceedings of the 2014 IEEE/RSJ International Conference on Intelligent Robots and Systems, Chicago, IL, USA, 14–18 September 2014; pp. 2120–2125.

25. Tatlicioglu, E.; Braganza, D.; Burg, T.C.; Dawson, D.M. Adaptive control of redundant robot manipulators with sub-task objectives. *Robotica* **2009**, *27*, 873–881. [CrossRef]

26. Merlet, J.P. *Parallel Robots*; Springer Science & Business Media: Berlin/Heidelberg, Germany, 2006; Volume 128.

27. Alici, G.; Shirinzadeh, B. Enhanced stiffness modeling, identification and characterization for robot manipulators. *IEEE Trans. Robot.* **2005**, *21*, 554–564. [CrossRef]

28. Gosselin, C. Stiffness mapping for parallel manipulators. *IEEE Trans. Robot. Autom.* **1990**, *6*, 377–382. [CrossRef]

29. Kim, H.S.; Tsai, L.W. Design optimization of a Cartesian parallel manipulator. *J. Mech. Des.* **2003**, *125*, 43–51. [CrossRef]

30. Görgülü, İ.; Kiper, G.; Dede, M.İ.C. A critical review of unpowered performance metrics of impedance-type haptic devices. In Proceedings of the European Conference on Mechanism Science, Aachen, Germany, 4–6 September 2018; pp. 129–136.

31. Salisbury, J.K.; Craig, J.J. Articulated hands: Force control and kinematic issues. *Int. J. Robot. Res.* **1982**, *1*, 4–17. [CrossRef]

32. Paul, R.P.; Stevenson, C.N. Kinematics of robot wrists. *Int. J. Robot. Res.* **1983**, *2*, 31–38. [CrossRef]

33. Kim, J.O.; Khosla, K. Dexterity measures for design and control of manipulators. In Proceedings of the IEEE/RSJ International Workshop on Intelligent Robots and Systems' 91 (Proceedings IROS'91), Osaka, Japan, 3–5 November 1991; pp. 758–763.

34. Yoshikawa, T. Manipulability of robotic mechanisms. *Int. J. Robot. Res.* **1985**, *4*, 3–9. [CrossRef]

35. Gosselin, C.; Angeles, J. A global performance index for the kinematic optimization of robotic manipulators. *J. Mech. Des.* **1991**, *113*, 220–226. [CrossRef]

36. Li, W.; Gao, F.; Zhang, J. R-CUBE, a decoupled parallel manipulator only with revolute joints. *Mech. Mach. Theory* **2005**, *40*, 467–473. [CrossRef]

37. Dhatt, G.; Lefrançois, E.; Touzot, G. *Finite Element Method*; John Wiley & Sons: Hoboken, NJ, USA, 2012.

38. Klimchik, A.; Chablat, D.; Pashkevich, A. Stiffness modeling for perfect and non-perfect parallel manipulators under internal and external loadings. *Mech. Mach. Theory* **2014**, *79*, 1–28. [CrossRef]

39. Deblaise, D.; Hernot, X.; Maurine, P. A systematic analytical method for PKM stiffness matrix calculation. In Proceedings of the 2006 IEEE International Conference on Robotics and Automation (ICRA 2006), Orlando, FL, USA, 15–19 May 2006; pp. 4213–4219.

40. Ghali, A.; Neville, A.; Brown, T.G. *Structural Analysis: A Unified Classical and Matrix Approach*; CRC Press: Boca Raton, FL, USA, 2014.

41. Pashkevich, A.; Chablat, D.; Wenger, P. Stiffness analysis of overconstrained parallel manipulators. *Mech. Mach. Theory* **2009**, *44*, 966–982. [CrossRef]

42. Wu, G.; Bai, S.; Kepler, J. Mobile platform center shift in spherical parallel manipulators with flexible limbs. *Mech. Mach. Theory* **2014**, *75*, 12–26. [CrossRef]

43. Hoevenaars, A.G.; Lambert, P.; Herder, J.L. Jacobian-based stiffness analysis method for parallel manipulators with non-redundant legs. *Proc. Inst. Mech. Eng. Part C J. Mech. Eng. Sci.* **2016**, *230*, 341–352. [CrossRef]

44. Klimchik, A. Enhanced Stiffness Modeling of Serial and Parallel Manipulators for Robotic-Based Processing of High Performance Materials. Ph.D. Thesis, Ecole des Mines de Nantes, Nantes, France, 2011.

45. Pashkevich, A.; Klimchik, A.; Chablat, D. Enhanced stiffness modeling of manipulators with passive joints. *Mech. Mach. Theory* **2011**, *46*, 662–679. [CrossRef]

46. Carbone, G. Stiffness analysis and experimental validation of robotic systems. *Front. Mech. Eng.* **2011**, *6*, 182–196.

47. Görgülü, İ.; Dede, M. Computation Time Efficient Stiffness Analysis of the Modified R-CUBE Mechanism. In Proceedings of the International Conference of IFToMM ITALY, Cassino, Italy, 29–30 November 2018; Springer: Berlin/Heidelberg, Germany, 2018; pp. 231–239.

48. Stocco, L.; Salcudean, S.; Sassani, F. Matrix normalization for optimal robot design. In Proceedings of the 1998 IEEE International Conference on Robotics and Automation, Leuven, Belgium, 16–20 May 1998; Volume 2, pp. 1346–1351.

49. Angeles, J. *Fundamentals of Robotic Mechanical Systems*; Springer: Berlin/Heidelberg, Germany, 2002; Volume 2.

50. Khan, W.A.; Angeles, J. The kinetostatic optimization of robotic manipulators: The inverse and the direct problems. *J. Mech. Des.* **2006**, *128*, 168–178. [CrossRef]

51. Cardou, P.; Bouchard, S.; Gosselin, C. Kinematic-sensitivity indices for dimensionally nonhomogeneous jacobian matrices. *IEEE Trans. Robot.* **2010**, *26*, 166–173. [CrossRef]

52. Cao, W.a.; Ding, H.; Zhu, W. Stiffness modeling of overconstrained parallel mechanisms under considering gravity and external payloads. *Mech. Mach. Theory* **2019**, *135*, 1–16. [CrossRef]

Article

Tailor-Made Hand Exoskeletons at the University of Florence: From Kinematics to Mechatronic Design [†]

Nicola Secciani [1,*], Matteo Bianchi [1], Alessandro Ridolfi [1], Federica Vannetti [2], Yary Volpe [1], Lapo Governi [1], Massimo Bianchini [3] and Benedetto Allotta [1,2]

[1] Department of Industrial Engineering, University of Florence, 50139 Florence, Italy; matteo.bianchi@unifi.it (M.B.); a.ridolfi@unifi.it (A.R.); yary.volpe@unifi.it (Y.V.); lapo.governi@unifi.it (L.G.); benedetto.allotta@unifi.it (B.A.)

[2] IRCCS Don Carlo Gnocchi Foundation, Via di Scandicci, 269, 50143 Florence, Italy; fvannetti@dongnocchi.it

[3] Institute for Complex Systems, National Research Council, Via Madonna del Piano, 10, 50019 Sesto Fiorentino, Italy; massimo.bianchini@isc.cnr.it

* Correspondence: nicola.secciani@unifi.it

† This paper is an extended version of our paper published in Secciani, N.; Bianchi, M.; Meschini, A.; Ridolfi, A.; Volpe, Y.; Governi, L.; Allotta, B. Assistive Hand Exoskeletons: The Prototypes Evolution at the University of Florence. In proceedings of the International Conference of IFToMM ITALY, Cassino, Italy, 29–30 November 2018.

Received: 13 March 2019; Accepted: 2 April 2019; Published: 3 April 2019

Abstract: Recently, robotics has increasingly become a companion for the human being and assisting physically impaired people with robotic devices is showing encouraging signs regarding the application of this largely investigated technology to the clinical field. As of today, however, exoskeleton design can still be considered a hurdle task and, even in modern robotics, aiding those patients who have lost or injured their limbs is surely one of the most challenging goal. In this framework, the research activity carried out by the Department of Industrial Engineering of the University of Florence concentrated on the development of portable, wearable and highly customizable hand exoskeletons to aid patients suffering from hand disabilities, and on the definition of patient-centered design strategies to tailor-made devices specifically developed on the different users' needs. Three hand exoskeletons versions will be presented in this paper proving the major taken steps in mechanical designing and controlling a compact and lightweight solution. The performance of the resulting systems has been tested in a real-use scenario. The obtained results have been satisfying, indicating that the derived solutions may constitute a valid alternative to existing hand exoskeletons so far studied in the rehabilitation and assistance fields.

Keywords: biomechanical engineering; wearable robotics; hand exoskeleton; mechanism design and optimization; kinematic analysis; mechatronics

---

1. Introduction

Over the past few decades, wearable robotics have been adopted in more and more sectors and, lately, the so called "assistive technology", that is the set of all the products that helps people to live as healthy, productive, independent, and dignified as possible, whatever their condition, has started to be more and more widely used also by the health care system [1–4]. There are more than 1 billion people all over the world who need at least one assistive device, however, and high costs and inadequate funding mechanisms allow only the 10% of the ones in need to have access to these products [5]. Keeping in mind the current state of the art [6–9], the authors have tackled this issue moving a step forward the democratization of the assistive technology by developing a low-cost hand exoskeleton to help and assist people with hand(s) impairments since, as easily verifiable, a key role in carrying out the Activities of Daily Living (ADLs) is played by the hands.

Robotic devices are thought to physically interact with human users suffering from disabilities for long periods of time [10] and, hence, they have to be designed meeting strict requirements in terms of safety, comfort and wearability. This is why one of the most difficult aspects of the human-robot interaction field is nowadays represented by the integration of robotics with assistive products. As if that were not enough, the complex anatomy and the wide variety of possible movements make the hand a great challenge both for the mechanical design and the control strategy [11].

An accurate state of the art assessment has been conducted, in the very first phase of the research activity, to define the underpinnings which the design process in based on, and, throughout the activity described in the paper, the critical evaluation of the wearable technologies in literature has been kept on to understand the research trends in designing exoskeletons responding to the patients' needs which, consequently, has paved the way to the development of an actually usable device.

An important aspect that must be considered is the clustering of the aimed technology. The design phase can be thus conducted heading to the fulfilling of each request group of the whole project (i.e., the exoskeleton design).

In accordance with the state of the art [6–9], hand exoskletons are classified using various criteria: linking system, Degrees of Freedom (DOFs) and actuation type.

As regards the linking system between the hand and the exoskeleton, there are two main different types: multi-phalanx devices [12,13], which directly control each phalanx separately, and single-phalanx exoskeletons [14], which actuate only that part of the hand they are connected to. The multi-phalanx approach exploits mechanisms made up of several parts and, thus, presents more complex control strategies [15–17]. Usually, these devices are not totally portable and they are supposed to be used for rehabilitative purposes [15,18] or in haptics [19], where the portability requirement is not a strict constraint. Nevertheless, this kind of devices allows to actuate the patients' hands exactly as well as they would do if they could by themselves. Single-phalanx devices use, instead, simpler actuation systems and control algorithms despite of less control capabilities than the multi-phalanx ones.

Another possible classification is based on the number of DOFs of the mechanisms. Rigid multi-DOFs kinematic chains are widely reported [20–22], while the number of rigid single-DOF mechanisms is not so large [23,24]. Since exoskeletons using a rigid multi-DOFs kinematic architecture demand multi-phalanx approaches, they usually present the same pros and cons. Current single-DOF devices present a very simplified kinematics [7,25], which is quite far from the physiological hand kinematics.

In recent years, soft-robotic applications have, then, increasingly been developed. They present a totally different type of mechanism based on elastomeric materials or fluid structures [26–30]. These devices result very lightweight and safe for the user because of their limited stiffness.

Concerning the type of actuator, hand exoskeletons may be driven by electric actuators [31,32] or pneumatic actuators [33]. The former actuation provides smaller forces to the hand than the latter, which, in turn, leads to higher weight and size due to its actuation system.

The proposed assistive and rehabilitative device for the hand focusing on the long fingers [34] has been designed considering the aforementioned research scenario. In particular, throughout the paper, three prototypes will be described to define the step by step design process that has lead to a novel single phalanx, rigid, single-DOF and cable-driven mechanism especially developed within this research activity.

In this manifold context, the use of optimization-based methods for the mechanical design, the exploitation of additive manufacturing technologies for the fabrication and considered choices of materials and electronics have proved to be effective tools for the development of well-performing prototypes of hand exoskeletons even in a low-cost perspective.

In this paper, the development process of a low-cost and fully wearable hand exoskeleton is discussed. The remainder of this section will present the overall framework this research activity has been carried out in. Then, starting from the same structure and kinematic architecture reported in Section 2), three different versions of the prototype have been sequentially developed to get closer to the user's needs. Sections 3–5 will describe the main accomplishments of each version in mechanical design, actuation system and control strategy.

Overall Framework

The research activity was conducted at the Mechatronics and Dynamic Modeling Laboratory (MDM Lab) of the Department of Industrial Engineering of Florence (DIEF). The MDM Lab has been active in the field of wearable robotics since 2013. In that year, the very first prototype started to be developed. A patient affected by Spinal Muscolar Atrophy (SMA) was the first user of the device, which was specifically developed for his needs and basing on his requirements. This first version of the hand exoskeleton prototype represented a first embodiment of the novel 1-DOF kinematic mechanism architecture which has then been later developed during the following years. In 2016, a collaboration with the Don Carlo Gnocchi Foundation Rehabilitation Center of Florence allowed to enlarge the target of possible users of the device. This scenario demands for the adaptation of the designed robotic system to different patients' hands. Exploiting the Motion Capture (MoCap) system available at the Don Gnocchi Rehabilitation Center, several studies focusing on the hand kinematics were carried out and a new Acrylonitrile Butadiene Styrene (ABS) exoskeleton was developed in accordance with the necessity of tailoring different fingers gestures. Currently, the collaboration with the Don Gnocchi Foundation deals with the study of innovative control strategies for hand exoskeleton systems based on surface ElectroMyoGraphic (sEMG) signals. Preliminary studies have been successfully concluded and some patients have already been enrolled for the testing campaign, which is about to start. At the time of writing, two projects are ongoing: HOLD, funded by the University of Florence and BMIFOCUS, funded by the Tuscany region.

2. Kinematic Architecture

Assistive robotic devices are, in general, made of both mechanical parts and electronics (e.g., sensors, power supply circuits, micro-processors and motors). They need thus to be carefully controlled in order to provide an intuitive and safe utilization. Achieving a smooth, comfortable, and robust control is a requirement that has to be kept in mind since the very beginning of the whole design process. The accurate development of a novel mechanism, characterized by a single DOF per finger allowed to precisely and comfortably reproduce the complex hand kinematics. Exploiting a single-DOF mechanism per finger granted for the control of only one variable (per finger) and resulted in the exploitation of less sensors and in the reduction of the computational burden.

The overall architecture of the system is split into two parts: a fixed frame, integral with the back of the hand, which houses motors and electronics, and four mobile finger mechanisms which act on the four long fingers. Motion and forces are transferred from the motors to the fingers by means of a cable transmission. An in-depth analysis of the kinematics of the single-DOF finger mechanism is presented in this paper for the first time and detailed in the following. For the sake of brevity, what reported below is related to just one finger mechanism, but the same analysis can be applied to all long fingers mechanisms as well.

Figure 1 shows the single-DOF kinematic chain exploited to move each long finger: The center of the reference system x_1y_1 related to the body A is fixed to the hand, roughly right above the MetaCarpoPhalangeal (MCP) joint. The other reference systems x_2y_2, x_3y_3, x_4y_4, x_5y_5 and x_6y_6 are integral with the bodies C, B, E, D and F. To simplify the notation each reference frame will also be related to a specific joint following the numerical progression (e.g., joint 1 is related to frame x_1y_1, joint 2 to frame x_2y_2 and so on). Component F is a thimble which has been added in the first version of the presented hand exoskeleton whose presence does not modify the 1-DOF kinematic chain of the device. For this reason, even if the thimble introduces a second connection point with the hand, this will not be considered a proper end-effector and the attention will mainly focus on component E.

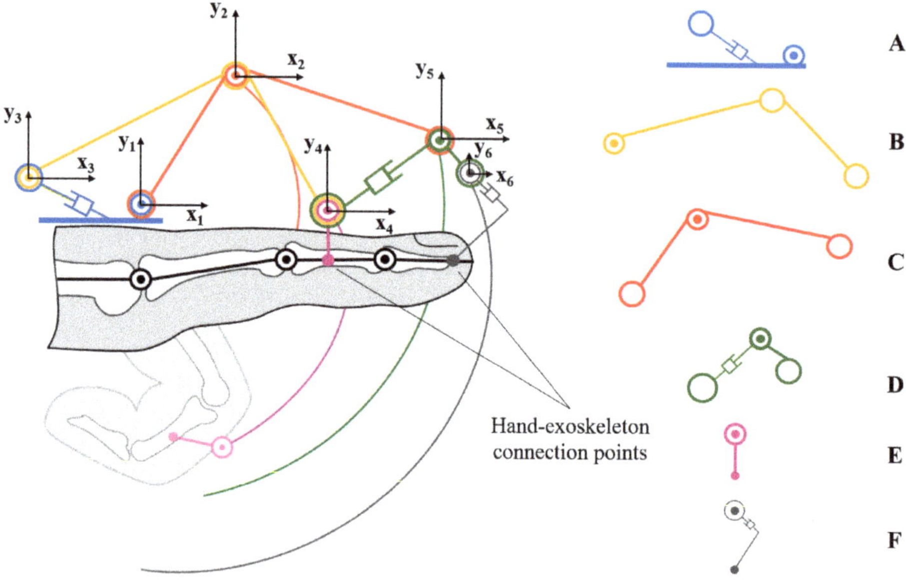

Figure 1. The figure shows the kinematic chain of the finger mechanism exploited in the presented work. While component A is integral with the back of the hand, the other parts are all mobile and their pose is uniquely identified once known the joint coordinate of joint 1.

The forward kinematics equations of the mechanism can be obtained starting from the revolute constraints, identifying rotational joints, in Ox_1y_1, Ox_2y_2, Ox_4y_4:

$$0 = {}^1\mathbf{p}_2 + \mathbf{R}_2^1 \, {}^2\mathbf{p}_1 \tag{1}$$

$$ {}^1\mathbf{p}_2 = {}^1\mathbf{p}_3 + \mathbf{R}_3^1 \, {}^3\mathbf{p}_2 \tag{2}$$

$$ {}^1\mathbf{p}_4 = {}^1\mathbf{p}_3 + \mathbf{R}_3^1 \, {}^3\mathbf{p}_4. \tag{3}$$

where, referring to Figure 1 and according to the mathematical notations reported in [6], the position of the origin of the i-frame with respect to j-frame has been denoted by the vector ${}^j\mathbf{p}_i = \left({}^jp_i^x \; {}^jp_i^y \right)^T \in \mathbb{R}^2$ (the component on $\mathbf{z}_i$ axis has been omitted as the proposed mechanism acts on a plane) and $\mathbf{R}_i^j$ represents the orientation of i-frame with respect to j-frame, which, in this case, results in a rotation about $\mathbf{z}_i$ axis through an angle α_i.

By analyzing the two cylindrical joints 3 and 5, the constraints equations are:

$$a_1 \, {}^1p_3^x + b_1 \, {}^1p_3^y + c_1 = 0 \tag{4}$$

$$a_2 \, {}^4p_5^x + b_2 \, {}^4p_5^y + c_2 = 0 \tag{5}$$

where

$$ {}^1\mathbf{p}_5 = {}^1\mathbf{p}_2 + \mathbf{R}_2^1 \, {}^2\mathbf{p}_5 \tag{6}$$

and

$$ {}^1\mathbf{p}_5 = {}^1\mathbf{p}_4 + \mathbf{R}_4^1 \, {}^4\mathbf{p}_5. \tag{7}$$

In Equations (4) and (5), a_1, b_1, c_1 and a_2, b_2, c_2 represent the two linear constraints of the mechanism and Equations (6) and (7) have been obtained considering the rotational joint in 5. Finally, even if it does not alter the kinematics of the device, an additional reference system (i.e., $x_6 y_6$) has been considered in the kinematic synthesis.

$$^1\mathbf{p}_6 = {}^1\mathbf{p}_4 + \mathbf{R}_4^1\, {}^4\mathbf{p}_6.$$

(8)

Referring to Equations (1)–(8), the state of the system is represented by the vector

$$\mathbf{q} = \left[{}^1\mathbf{p}_2^T, {}^1\mathbf{p}_3^T, {}^1\mathbf{p}_4^T, {}^1\mathbf{p}_5^T, {}^1\mathbf{p}_6^T, \alpha_2, \alpha_3, \alpha_4 \right]^T \in \mathbb{R}^{13}$$

(9)

and depends on the control variable α_2. The unknowns representing the state of the system can be thus calculated as a function of only α_2 by solving Equations (1)–(8). All the interesting points of the mechanism (included in the state vector $\mathbf{q}$) are in fact completely described as functions of the angle α_2 and of the geometrical parameters $\mathbf{S} \in \mathbb{R}^{16}$:

$$\mathbf{S} = [{}^2\mathbf{p}_1^T, {}^3\mathbf{p}_2^T, {}^2\mathbf{p}_5^T, {}^3\mathbf{p}_4^T, {}^4\mathbf{p}_6^T, a_1, b_1, c_1, a_2, b_2, c_2]^T.$$

(10)

All these parameters are completely known because they represent geometric quantities, depending only on the design of the exoskeleton parts. Consequently, it is possible to solve the extended direct kinematic model $\tilde{\mathbf{q}} = \mathbf{f}(\alpha_2, \mathbf{S}) \in \mathbb{R}^{12}$ (see Equation (11)) of the mechanism writing a function of α_2 and $\mathbf{S}$, where $\tilde{\mathbf{q}}$ is the unknown part of the state vector $\mathbf{q}$:

$$\tilde{\mathbf{q}} = [{}^1p_2^x, {}^1p_2^y, {}^1p_3^x, {}^1p_3^y, \alpha_3, {}^1p_4^x, {}^1p_4^y, \alpha_4, {}^1p_5^x, {}^1p_5^y, {}^1p_6^x, {}^1p_6^y]^T = \mathbf{f}(\alpha_2, \mathbf{S}).$$

(11)

The closed form resolution of the aforementioned forward kinematic is given hereinafter. Each component of vector $\tilde{\mathbf{q}}$ is highlighted in blue when it is solved in terms of only α_2 and elements of $\mathbf{S}$. Starting from Equation (1), it is possible to obtain Equations (12) and (13):

$$^1p_2^x(\alpha_2) = -\left(c\alpha_2 \cdot {}^2p_1^x - s\alpha_2 \cdot {}^2p_1^y \right)$$

(12)

$$^1p_2^y(\alpha_2) = -c\alpha_2 \cdot {}^2p_1^y - s\alpha_2 \cdot {}^2p_1^x.$$

(13)

From Equations (2) and (4), the following equations can be written:

$$\begin{cases} {}^1p_2^x - {}^1p_3^x = c\alpha_3 \cdot {}^3p_2^x - s\alpha_3 \cdot {}^3p_2^y \\ {}^1p_2^y - {}^1p_3^y = c\alpha_3 \cdot {}^3p_2^y + s\alpha_3 \cdot {}^3p_2^x \\ a_1 \cdot {}^1p_3^x + b_1 \cdot {}^1p_3^y + c_1 = 0. \end{cases}$$

(14)

and solving the system:

$$^1p_3^y(\alpha_2) = \frac{-\left({}^1p_2^x \cdot \frac{b_1}{a_1} + \frac{b_1 \cdot c_1}{a_1{}^2} - {}^1p_2^y \right)}{\left(\frac{b_1}{a_1} \right)^2 + 1} + \frac{\sqrt{\left({}^1p_2^x \cdot \frac{b_1}{a_1} + \frac{b_1 \cdot c_1}{a_1{}^2} - {}^1p_2^y \right)^2 - \left[\left(\frac{b_1}{a_1} \right)^2 + 1 \right] \cdot H}}{\left(\frac{b_1}{a_1} \right)^2 + 1}$$

(15)

$$^1p_3^x(\alpha_2) = -\frac{1}{a_1} \cdot \left(b_1 \cdot {}^1p_3^y + c_1 \right)$$

(16)

where

$$H = \left({}^1p_2^x \right)^2 + \left({}^1p_2^y \right)^2 - \left({}^3p_2^x \right)^2 - \left({}^3p_2^y \right)^2 + 2 \cdot {}^1p_2^x \cdot \frac{b_1}{a_1} + \left(\frac{c_1}{a_1} \right)^2.$$

(17)

and depends only on known values. Now α_3 can be computed as:

$$\alpha_3(\alpha_2) = a\tan 2 \left(\frac{\left(-{}^1p_2^x + {}^1p_3^x + \frac{{}^3p_2^x}{{}^3p_2^y}\cdot{}^1p_2^y - \frac{{}^3p_2^x}{{}^3p_2^y}\cdot{}^1p_3^y\right)}{{}^3p_2^y + \frac{\left({}^3p_2^x\right)^2}{{}^3p_2^y}} , \frac{{}^1p_2^y - s\alpha_3\cdot{}^3p_2^x - {}^1p_3^y}{{}^3p_2^y} \right). \tag{18}$$

At this point, Equations (19) and (20) can be written from Equation (3):

$$ {}^1p_4^x(\alpha_2) = {}^1p_3^x + c\alpha_3\cdot{}^3p_4^x - s\alpha_3\cdot{}^3p_4^y \tag{19}$$

$$ {}^1p_4^y(\alpha_2) = {}^1p_3^y + c\alpha_3\cdot{}^3p_4^y + s\alpha_3\cdot{}^3p_4^x. \tag{20}$$

From Equation (6), one can get

$$ {}^1p_5^x(\alpha_2) = {}^1p_2^x + c\alpha_2\cdot{}^2p_5^x - s\alpha_2\cdot{}^2p_5^y \tag{21}$$

$$ {}^1p_5^y(\alpha_2) = {}^1p_2^y + c\alpha_2\cdot{}^2p_5^y + s\alpha_2\cdot{}^2p_5^x \tag{22}$$

and from Equations (5) and (7):

$$\begin{cases} {}^1p_5^x - {}^1p_4^x = c\alpha_4\cdot{}^4p_5^x - s\alpha_4\cdot{}^4p_5^y \\ {}^1p_5^y - {}^1p_4^y = c\alpha_4\cdot{}^4p_5^y + s\alpha_4\cdot{}^4p_5^x \\ a_2\cdot{}^4p_5^x + b_2\cdot{}^4p_5^y + c_2 = 0. \end{cases} \tag{23}$$

As in the previous cases, it can be obtained

$$ {}^4p_5^y(\alpha_2) = \frac{-(b_2\cdot c_2) + \sqrt{(b_2\cdot c_2)^2 - \left(b_2{}^2 + a_2{}^2\right)\cdot T}}{b_2{}^2 + a_2{}^2} \tag{24}$$

$$ {}^4p_5^x(\alpha_2) = -\frac{1}{a_2}\cdot\left(b_2\cdot{}^4p_5^y + c_2\right) \tag{25}$$

where:

$$ T = -a_2{}^2\cdot\left[\left({}^1p_5^x - {}^1p_4^x\right)^2 + \left({}^1p_5^y - {}^1p_4^y\right)^2\right] + c_2{}^2. \tag{26}$$

α_4 is now calculated as:

$$\alpha_4(\alpha_2) = atan2 \left(\frac{\left(-{}^1p_4^y + {}^1p_5^y - \frac{{}^4p_5^y}{{}^4p_5^x}\cdot{}^1p_5^x + \frac{{}^4p_5^y}{{}^4p_5^x}\cdot{}^1p_4^x\right)}{-{}^4p_5^x + \frac{\left({}^4p_5^y\right)^2}{{}^4p_5^x}} , \frac{{}^1p_5^x + s\alpha_4\cdot{}^4p_5^y - {}^1p_4^x}{{}^4p_5^x} \right). \tag{27}$$

Finally, ${}^1\mathbf{p}_6$ results:

$$ {}^1p_6^x(\alpha_2) = {}^1p_4^x + c\alpha_4\cdot{}^4p_6^x - s\alpha_4\cdot{}^4p_6^y \tag{28}$$

$$ {}^1p_6^y(\alpha_2) = {}^1p_4^y + c\alpha_4\cdot{}^4p_6^y + s\alpha_4\cdot{}^4p_6^x \tag{29}$$

At this point, the motion is completely described and the positions of every joint relative to the 1-frame is given. Figure 1 also shows the trajectories of the finger mechanism joints, providing a qualitative overview of the resulted kinematics of the mechanism when fingers are actuated.

3. First Prototype: Kinematic Validation

A first version of the hand exoskeleton (shown Figure 2 and detailed in [34,35]) has been designed and manufactured to test the embodiment of the kinematic model proposed in Section 2. A patient affected by SMA was the first user of the device, specifically developed for him.

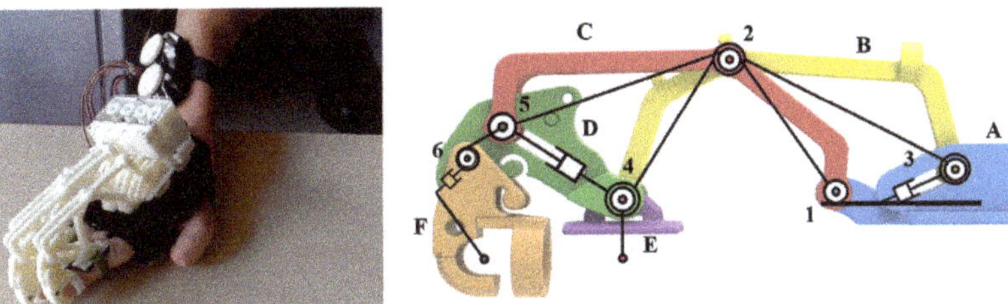

Figure 2. The figure shows, on the left, the first version of the hand exoskeleton prototype worn by the patient and, on the right, the corresponding kinematic chain and Computer Aided Manufacturing (CAD) model. Colors and names (capital letters) of the components, and joints enumeration are reported as introduced in Section 2.

3.1. Mechanical Design

All the mechanical parts have been 3D-printed using a Dimension Elite by Stratasys in ABS thermoplastic polymer. One of the main advantages of 3D-printing technology is that it allows to manufacture components without considering technological constraints due to the particular production method as well as it may happen using subtracting processes. Then, among different 3D-printable materials, ABS has been chosen because it represents a satisfying trade-off between good mechanical characteristics, lightness, and cheapness.

The design process was composed of three sequential phases. Firstly, 2D trajectories of the index finger of the specific user were acquired exploiting open source software Kinovea, https://www.kinovea.org/ (accessed on 26 March 2019). Secondly, an optimization MATLAB-based algorithm [36] minimized the constrained nonlinear multi-variable function describing the kinematics of the mechanism, modifying its geometrical parameters and leading the mechanism kinematics to fit the acquired trajectories. Thirdly, a scaling phase allowed to adapt the kinematic model to each patient's finger by resizing the geometry of the index finger mechanism accordingly to the dimensions of the other fingers. Then, once the mechanism features was defined, virtual tests exploiting SolidWorks Motion Simulation tools have then been carried out to assess, before the manufacturing of the device, the hand-exoskeleton kinematic coupling and interaction in simple opening and closing gestures. Finally, right before the manufacturing, the mechanical parts have undergone a further reshaping process which has changed their structure, but which has not changed their overall kinematics (as shown in Figure 2, the shapes of the real parts differ from the straight lines of the kinematic model, but the position of the joints has remained unchanged). This later step was required to avoid possible interpenetrations with the hand and, as described in the following lines, for reasons of mounting and loading conditions.

In particular, component C was split in two parts guaranteeing a symmetric load configuration during the use of the exoskeleton and obtaining a more stable solution. Components D was made in two parts as well, allowing them to be assembled together. Component E, which represents the hand-exoskeleton interface, has been designed to wrap only the back side of the finger phalanx not to reduce the sense of touch, while a Velcro held the finger tight achieving a solid connection. To reduce the lateral encumbrances of each mechanism, pins and shafts were directly integrated in the ABS components as lateral rods. All the aforementioned modifications are graphically reported in Figure 3.

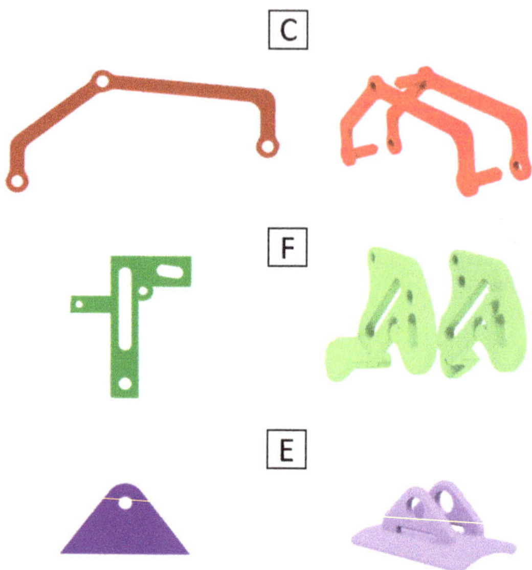

Figure 3. The figure shows the adaptation of components C, F, and E.

3.2. Actuation System and Control Strategy

As visible in Figure 2, both the transmission and the actuation system are placed on the hand backside, as well as the mechanisms are positioned on the fingers, and they do not impede objects handling. The reduction of the total mass, which was one of the main requirements of the device, has led to the choice of high power density actuators to be directly mounted on the back of the hand.

Four Savox SH-0254 servomotors, http://www.savoxusa.com (accessed on 26 March 2019), one per long finger, have been selected for their characteristics: maximum torque of 0.38 Nm and maximum angular speed of 7.69 rad/s at 6.0 V, with a size of 22.8 × 12 × 29.4 mm and weight of 16 g. These motors have been modified to allow for the continuous rotation of the shaft despite the resulting loss of position feedbacks. The four servomotors are in charge of opening the fingers at the same time by pulling cables which have two connection points on each finger mechanism. Closing gesture is, instead, passively allowed releasing the same cables.

The control unit was based on a 6-channels MicroMaestro control board, https://www.pololu.com (accessed 26 March 2019) which has been chosen for its cheapness, its lightness (only 5 g), its small dimensions (21 × 30 mm) and, above all, because its six channel matched the number of external devices that had to be connected to the board: the aforementioned four servomotors and two buttons, one for opening and one for closure triggering action. The control unit and the actuators were powered by a compact 4-cell Lithium battery (at 6.0 V), which was placed in an elastic band on the arm of the user, provided with a safety switch close to the buttons case, mounted instead on the forearm.

Regarding the control strategy, the system was controlled by a simple script, stored and running directly on the MicroMaestro chip-set. The code had to continuously check for one of the two buttons to be pushed and held down and then react by sending the corresponding command to the actuators. This version did not include sensors for fingers position feedback and the bounds of the exoskeleton range of motion were manually managed by the user (keeping pushed or releasing the buttons) in order not to overcome his anatomical limits. Even thought there was one motor per finger, the possibility to move each of them independently from the others has not been considered for simplicity and all the long fingers were moved together.

3.3. Testing

The whole mechatronic system described in the previous sections has then been assembled and tested to evaluate the manufacturability of the adopted solution (see Figure 2). The single DOF mechanism, the one visible in Figure 1, consists in a closed kinematic chain mechanism capable of applying a defined roto-translation motion to the middle point of the intermediate phalanx. Again, Kinovea software has been employed in this testing phase. The exoskeleton has been worn by the user who has been told to perform some opening and closing gestures. Colored markers have been placed on the mechanism joints and their trajectories have been extracted from the images. Finally, the acquired trajectories has been compared to the kinematic model ones. The comparison between real data and simulated trajectories are presented in Figure 4 only for the index finger mechanism. The same considerations can be adopted for all other long fingers.

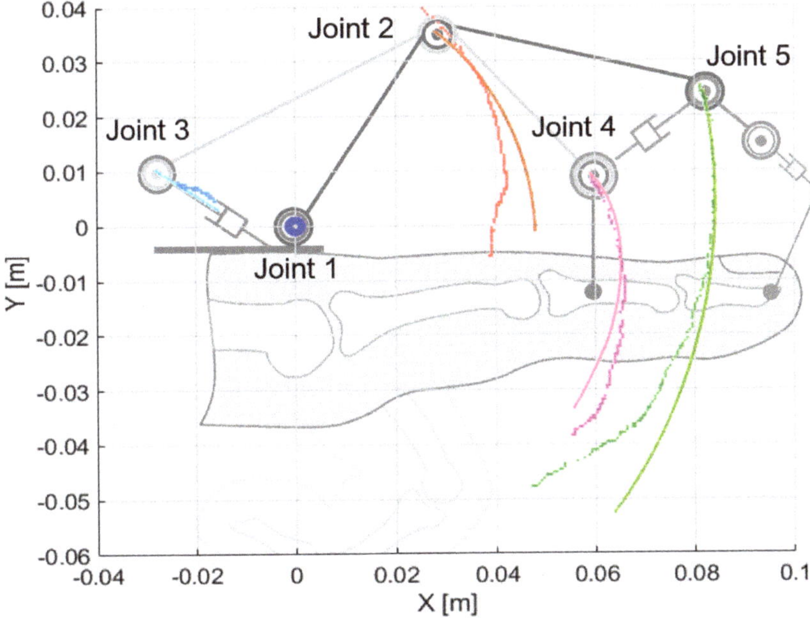

Figure 4. The figure shows the comparison between the trajectories of the joints of the exoskeleton generated by the kinematic model (solid lines) and the real ones (dashed lines) for the index finger.

Despite the fact that the comparison of the trajectories has given encouraging results—in particular with regard to the joint 2, which is the main point of interaction between the exoskeleton and the finger—tests conducted on this first version highlighted several flaws which negatively impacted on its usability.

- The patient pointed out that being forced by the exoskeleton to move the fingers only on the flex/extension plane (this prototype did not allow for the ab/adduction movement of the MCP joint) resulted in an extremely annoying feeling.
- As well as the comfort, also the intuitiveness of control was very low: the patient had, in fact, to use both hand to move just one because of the buttons-driven actuation system.
- The lack of a proper position feedback over the flex/extension angle of the finger did not allow for a safe automatic control over the Range of Motion (ROM).

- The solution provided by the exploited optimization algorithm was, by the nature of the algorithm itself, strongly dependable on the choice of the initial state, which was arbitrarily made accordingly to the imposed geometric constraints. This process hence resulted in being low-adaptable to different hand sizes because of the necessity of proper setting the initial state of the algorithm every time, and it also offered no guarantees of finding a real optimum solution.

The tested device hence resulted in being still far from a clinical application and many improvements had to be made to move towards that direction.

4. Second Prototype: Ergonomics and Adaptability Improvements

Starting from the weaknesses of the exoskeleton version discussed in Section 3 and considering that the adaptability and the ease of wearing requirements must be, not only kept, but also improved, a second prototype of the hand exoskeleton has been developed and printed. In particular, thanks to the experience gained during the testing phase several modifications have been made to produce a more comfortable and wearable system, and develop a new design procedure easier to adapt to different users. The high wearability of a system is hence endorsed by its transparency with respect to the hand natural kinematics, rising from a device which is not felt as a constraint, but whose use actually results straightforwardly intuitive. Both the optimization strategy, leading to a mechanism which replicates at best the fingers gestures, and the revamping design, heading to a lighter solution, contribute to a system better accepted by the end user.

4.1. Mechanical Design

The optimization algorithm, in charge of modifying the geometry of the device accordingly to the fingers trajectories has been replaced: a Nelder-Mead based optimization algorithm, used to solve non convex, non linear constrained problems, has been applied achieving a straightforward adaptability to several users [37]. Taking acquisition data (collected exploiting a BTS SMART-Suite MoCap System by BTS Bioengineering (BTS Bioengineering, Garbagnate Milanese (MI), Italy)) and the kinematics of the mechanism as inputs, the implemented algorithm provides a customized geometry specific for each patient.

A CAD model has then been developed by using parametric dimensions to make the whole procedure automatic. This choice leads to two important consequences. Firstly, the position of each joint results directly connected to the outputs of the optimization routine without requiring manual intervention. In addition, as well as tested with the first prototype, a parametric CAD model enable wearing simulations. By replicating virtual opening and closing gestures with the device worn by the user, the exoskeleton kinematics and its coupling with each finger can be checked and assessed. The ABS prototype is then print out exploiting the Fused Deposition Modeling (FDM) additive manufacturing technique.

Even though the overall kinematics of the finger mechanism has not been changed with respect to the first prototype, the mechanical architecture of the exoskeleton has been subjected to important modifications. The thimble wrapping the finger tip has been eliminated leaving only one contact point between each finger and the device. By removing this component, the driving purpose of the finger mechanism has been kept but the uncomfortable feeling of having the whole finger constraint to a rigid system has been avoid. This choice also leaves the touching feeling while grabbing objects. A passive DOF has been added upstream joint 1 (refer to Figure 1) to follow finger abduction and adduction motions during hand opening and closing. This solution also improves the auto-alignment between finger and mechanism joints.

4.2. Actuation System and Control Strategy

To get closer to the needs come out during the tests on the previous prototype, this new "release" of the hand exoskeleton presents only two servomotors Hitec HS-5495BH (http://hitecrcd.com/products/servos/sport-servos/digital-sport-servos/hs-5495bh-hv-digital-karbonite-gear-sport-servo/product): one of them guides the index finger flexion/extension, the other one is in charge of moving the other three long fingers. In addition to reducing the overall weight and encumbrance of the system, reducing the number of actuators overcome an issue that has immediately occurred the high difficulty to prevent the four motors from acquiring more and more relative phase shift with the prolonged use.

Two motors, one the other side, require an important change in the transmission system: because of the different sizes of each finger mechanisms (due to, in turn, the dissimilar dimensions of the fingers), closing and opening velocity for each finger must be different one to the others. To overcome this issue, the pulley spliced to the output shaft of the actuator in charge of moving three long fingers has been thought with three different diameters. That introduced a set gear ratio and, since the wires connected to the three fingers mechanisms winds pulley with different diameters, middle, ring and small finger mechanisms are moved at the same time with the same motor but each one at its own suitable speed. The overall weight of this new prototype resulted in 242 g. Regarding the characteristics of the new actuators, they present a maximum torque of 0.735 Nm at 7.4 V with a size of 39.8 × 19.8 × 38.0 mm and a weight of 44.5 g. The maximum angular speed is 66.67 rad/s at 7.4 V.

Also the control system has been developed aiming to the same goal headed to during the mechanical design: the total costs, complexity and weight of the system must be kept as low as possible. In this framework, Arduino Nano represented a simple, cheap, but performing solution. Two magnetic encoders (15-bit resolution) are spliced on joint 1 of the finger mechanisms (Figure 5). Since one servomotor actuates the index and the other one the other three long fingers, in order to guarantee a suitable control of the grasping, only two encoders are needed: one placed on the index and one on the little finger. They measure the value of the angle α_2, which identifies pose of the mechanism and, consequently, of the finger fixed to it.

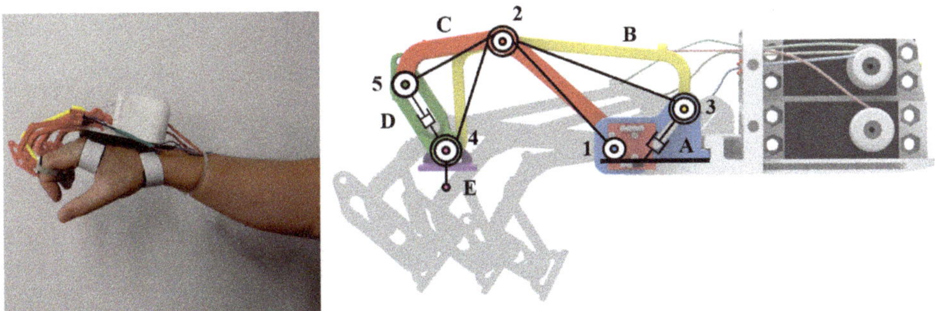

Figure 5. The figure shows, on the left, the second version of the exoskeleton system worn by a healthy subject and, on the right, the corresponding kinematic chain and CAD model. Colors and names (capital letters) of the components, and joints enumeration are reported as introduced in Section 2.

The control strategy is characterized by two main control loops that have been added to safely stop the motion of the exoskeleton once the ROM limits are reached (namely maximum opening and closure), and to check if an object is grabbed. While the permanence within the ROM boundaries is easily checked by means of the direct measurement from the encoders, the grabbing of an object is recognized and detected by the evaluation of the length of the unrolled cable twice. The first evaluation is made by the kinematic equations of the mechanism and the second one exploiting the motors speed and the pulleys radius. Comparing the differential measurement to a fixed threshold (set at 10 mm, which is roughly the length of a quarter of the motor pulley circumference) allows to understand if the actuation system is still releasing cable while the hand is not further moving. This means that an object is likely grasped. In both cases each motor stops while holding its current angular position.

Testing

In this the results obtained related to the validation and testing of the second prototype will be presented and discussed. A real ABS prototype has been hence 3D-printed to test the manufacturability of the kinematic architecture and assess the goodness of the optimization strategy for different users. The tests have been carried out at the Don Carlo Gnocchi Foundation Rehabilitation Center in Florence, Italy, making use of the MoCap system used also to evaluate the motion of the hand alone; the cameras configuration exploited to validate the second prototype is the same used during the motion capture of the hand.

The right index finger motion of the 13 subjects enrolled to validate the performance of the optimization procedure has been assessed through a MoCap analysis during three consecutive flexion/extension movements. Taking the aforementioned trajectories as reference, the proposed optimization process has been carried out over the 13 healthy subjects and the results are reported in Table 1.

Table 1. Length and width of the hand, maximum, average and standard deviation (Std Dev) of the error between the desired and the real trajectory for the index finger, and maximum, average and standard percentage error expressed with respect to the subjects' finger length.

Dimensions [mm^2]	Error [mm]			Error/Finger Extension [%]		
Length $\times$ Width	Maximum	Average	Std Dev	Maximum	Average	Std Dev
200 $\times$ 94	4.96	2.66	1.18	4.5	2.4	1.1
165 $\times$ 82	3.72	2.29	1.13	4.2	2.6	1.3
191 $\times$ 78	4.82	3.15	1.30	4.7	3.1	1.3
192 $\times$ 84	5.59	3.68	1.70	5.5	3.6	1.7
189 $\times$ 95	5.92	3.83	1.63	5.9	3.8	1.6
192 $\times$ 91	3.05	2.05	0.81	3.0	2.0	0.8
197 $\times$ 88	3.45	2.10	1.21	3.3	2.0	1.1
193 $\times$ 89	6.30	4.12	1.40	6.3	4.1	1.4
187 $\times$ 81	0.90	0.53	0.31	0.9	0.5	0.3
150 $\times$ 75	4.76	2.92	1.42	5.9	3.6	1.7
193 $\times$ 80	4.99	3.08	1.48	4.9	3.0	1.4
220 $\times$ 90	8.83	6.80	1.79	7.6	5.9	1.5
195 $\times$ 92	5.57	3.96	1.24	5.3	3.8	1.2

The results reported in Table 1 allowed to verify and check the reliability of the optimization algorithm when different hand exoskeletons have to be adapted to hands characterized by different dimensions. The average of the maximum calculated errors (this measurement provides a quantitative value of the difference—maximum distance—between the trajectory tracked by the user's finger and the one generated by the finger mechanism of hand exoskeleton) was 3.16 mm (standard deviation 1.47 mm). That proves the goodness of the implemented optimization strategy in adapting the a-priori defined one-DOF kinematic chain to the user's anatomy.

These new tests have shown promising results with regard to the portability of the device. With the addition of the passive DOF on the ab/adduction movement of the MCP joint, the exoskeleton is considerably more compliant with the hand movements and can be worn longer without feeling as uncomfortable as before. Moreover, the automatic control over the ROM reliably prevents the exoskeleton to move the fingers towards unnatural or painful positions, liberating the user from the fear of hurting himself if not enough attention was paid. The result achieved with new optimization strategy of the geometry of the finger mechanism can be considered even more important though. In fact, this new process allowed the mechanism to be quickly adapted to different users, paving the way to future experimental campaigns on multiple patients. However, the buttons-triggered actuation still remained an open up point to be further investigated.

5. Third Prototype: Intuitive Control

This current version of the exoskeleton (Figure 6) is represented by a fully portable, wearable and highly customizable device that can be used both as an assistive hand exoskeleton and as a rehabilitative one. Both mechatronic design and control system are developed basing on the patients needs in order to satisfy users' daily requirements increasing their social interaction capabilities.

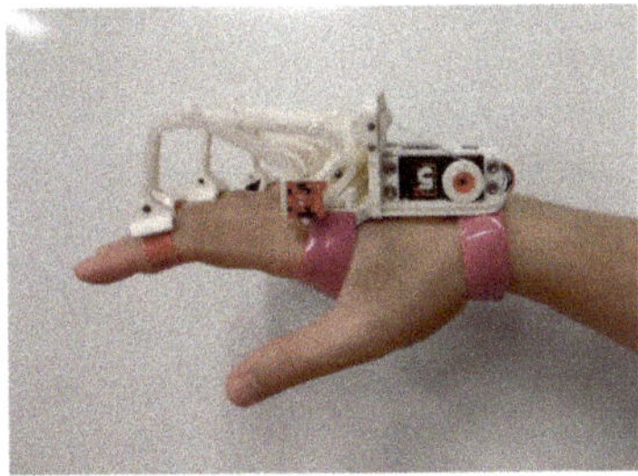
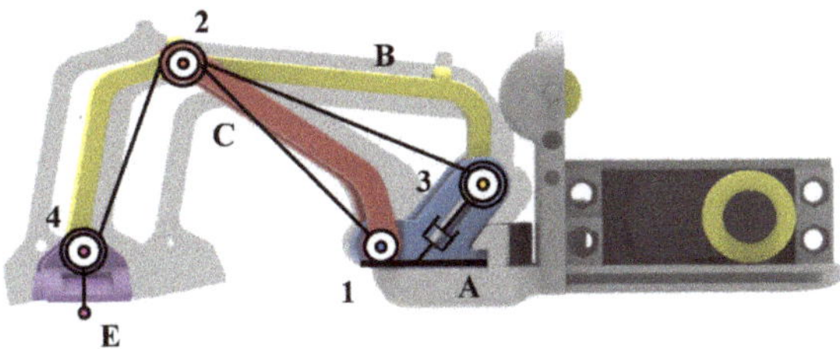

Figure 6. The figure shows, on the left, the final and current version of the exoskeleton prototype developed by the DIEF worn by a healthy subject and, on the right, the corresponding kinematic chain and CAD model. Colors and names (capital letters) of the components, and joints enumeration are reported as introduced in Section 2.

5.1. Mechanical Design

The mechanics of this last exoskeleton has been revamped to achieve a more lightweight solution improving its wearability without influencing the obtained results in terms of accuracy while replicating hand gestures. The new system is actuated by a single servomotor and a specific cable driven transmission system has been developed to open all the four long fingers together at the same time. Different mechanisms velocities are obtained thanks to different pulleys diameters, which are calculated depending on users' fingers dimensions, and coupled with the same common shaft which is moved by the motor through a belt transmission. The mechanism kinematic architecture has been further modified by eliminating component D of Figure 1. Thickness of components C and B has thus been increased to bear the heavier load cycle these components are now subjected to during functioning.

5.2. Actuation System and Control Strategy

The first important difference with respect to the previous systems is that, as reported above, another motor has been removed adopting a single-motor actuation system. Even though the motor has not been changed, the exploitation of just one actuator has brought with it some advantages: the total weight of the system has been remarkably reduced and the control code has resulted in being computationally lighter, not having to manage the coordination between motors. Nevertheless, the main difference of this prototype lies in the triggering system. Tests conducted on the second version of the prototype have hence stressed the importance for the user of being able to use both hands independently, pushing the buttons-based triggering action to be replaced with something which could allow for an autonomous control of each hand. Following the most recent research trends in literature (e.g., [38,39]), a specific EMG-based control architecture has been developed letting the user the total control of the exoskeleton actuation without being forced to use the other hand (as it happened with the first and the second prototypes).

Accordingly to the guidelines of the research work, also the electronics of the system has been reduced to the minimum necessary. Two MyoWare Muscle Sensors (AT-04-001) by Advancer Technologies https://learn.sparkfun.com/tutorials/myoware-muscle-sensor-kit have been chosen for collecting EMG signals from specific forearm muscles, i.e., flexor digitorum superficialis and extensor digitorum superficialis. These sensors, small (20.8 × 52.3 mm) and low-powered, measure the electrical activity of a muscle, outputting either raw EMG signals or enveloped EMG signals, which are amplified, rectified and integrated.

The proposed control strategy, presented in [40] and detailed in [41], can be split in two main parts, which are sequentially executed every 33 ms (i.e., at 30 Hz). The first one is in charge of classifying user's intentions from the muscular activity measurements, the second one instead manages the corresponding actuation of the system. Within the second part of the code, an outer control loop checks if the system overcomes a fixed ROM and an inner control loop, which is only active during hand closing, is meant to check if an object is grasped. by evaluating the closing velocity of the index finger. when it drops below a fixed threshold while the motor is still running, it is reasonable to think that the hand has encountered an object or an obstacle. Real-time information about the position and the velocity of the index finger is collected by means of the magnetic encoder mounted on the exoskeleton in correspondence of the MCP joint of the finger.

Since the human hand can perform lots of different gestures and the corresponding muscles are very close to each other, a precise classification of every user intention usually requires the use of workstations, which is definitely far away from the idea of cheapness and wearability this project is based on. Hand opening, hand closing and hand resting have then been considered as the only possible user's intentions to be classified, as they represent the basic hand motions for the ADLs.

The classification phase is carried out by means of an algorithm called "Point-in-Polygon algorithm". This algorithm works on the base of a ray-casting to the right. It takes as inputs the number of the polygon vertices, their coordinates and the coordinates of a test point; each iteration of the loop, the line (ray) drawn (cast) rightwards from the test point is checked against the polygon perimeter and the number of times this line intersect a different edge is counted; if the number of crosses is an odd number of times, then the point is inside, if an even number, the point is outside. This classifier is tuned during a preliminary training phase through a custom Qt Graphical User Interface (GUI) developed by the authors. It is a user-friendly tool which allows for collecting EMG signals and for displaying them on a 2D Cartesian plane, whose axis report respectively data from the first and the second sensor. Once the EMG data has been collected for all the three aforementioned gestures, it is possible to manually draw the geometric figures which delimit the data corresponding to the same gesture. The correct choice of the parameters of the polygons (e.g., vertices, shape, size) represent a crucial point of the classification phase which is supposed to be performed basing on patient's needs to improve the accuracy and the rejection of disturbs.

5.3. Testing

At the time of writing, the version of the exoskeleton presented in this section is at the heart of an application for an experimental campaign protocol that has not been submitted to an Ethical Committee yet. Only a few preliminary tests have been conducted and their results can be found in [41]. Although quantitative data about the proposed control strategy still lack, it can be asserted that it, surely, represent a step towards intuitiveness and a step forward for the prototype itself.

6. Conclusions

This work collects the results of the research activity on wearable robotics carried out at the Mechatronics and Dynamic Modeling Laboratory of the Department of Industrial Engineering of the University of Florence during the years 2015–2018. The activity focused on the study of design strategies for hand exoskeletons, starting from the analysis of state-of-the-art solutions and leading to the development of new robotic devices. Hand exoskeletons mechanical design constituted the main subject of the research, motivated by the issue that people who have lost or injured hand skills are deeply compromised in the possibility of an independent and healthy life with a sensitive reduction in the quality of their lives themselves. Since robotic devices, such as hand exoskeletons, may represent an effective solution for the patient affected by hand opening disabilities, or other mobility impairments, the focus was particularly given to the development of a robotic solution provider of an aid during prolonged and high-intensity rehabilitation treatments. Such systems also contribute to reduce costs and burden for the therapists, as long as being an effective aid in the execution of ADLs.

The overall research activity aimed not only to present the design of the hand exoskeletons developed by the University of Florence, but also, and actually mainly, to propose new user-centered and patient-based strategies which can be adopted in the production of wearable systems. The optimization-based scaling strategy (which, starting from the users' own gestures, leads to a tailor-made device) along with the mechanical choices (which guarantee the outline of a lightweight and compact overall system) center the end-user on the design process and the proposed wearable systems are fully designed around him. In all the presented hand exoskeleton prototypes, the system kinematics plays the main role for the design of a tool as much wearable as possible, which guarantees a comfortable feeling even during the prolonged use.

Starting from the same kinematic architecture, three different hand exoskeletons have been developed heading for four features mandatory for the actual use of the device: the comfortable feeling for the user, the adaptability of the kinematic chain of the finger mechanism to the patient's hand, the safe use of the device, and the user-friendly activation system and control of the exoskeleton. Table 2 summarizes the main results achieved in the design of each prototype.

Table 2. Results achieved in the design of the three developed prototypes. Symbol ✓ indicates the achievement of an acceptable level for the feature reported on the first column of the table, while symbol ✗ points out an open up point that had to be revised in the next version.

Performance Criteria	Prototype		
	1	2	3
Number of DOFs	1	1	1
Number of linkages	6	5	4
Trigger strategy	Buttons	Buttons	EMG-based
Auto-alignment	✗	✓	✓
Comfort for the user	✗	✓	✓
Adaptability to the hand	✗	✓	✓
Safe use	✗	✓	✓
User-friendly activation	✗	✗	✓

However, achieving the aforementioned feature was particularly difficult when the constraints of the single DOF had to be maintained. The proposed solution resulted in a tight (due the 1 DOF per finger, cable-driven, single-motor system) device capable of precisely replicating the hand gestures. The satisfying results highlighted the goodness of the derived solutions, which may constitute a suitable alternative to existing hand exoskeletons. Nonetheless, there is still room for improvement: for instance, clinical trials exploiting the reached device may allow for the assessment of the impact the hand exoskeleton would have during real rehabilitation sessions. A testing phase, employing a real actuation system, represents the first opened up point, which could lead to the necessity of improving the mechanical architecture of the whole exoskeleton in order to increase its usability and efficacy. Also the employment of different materials, whose structural features may benefit the robustness of the device, inspires the possibility for further investigations [42,43]. The last, but not least, important issue which is worth being explored in the near future is the assessment of the pose of the hand depending on the task it is required to accomplish. As reported in [44], the control of hand posture involves a few postural synergies leading to reduce the number of degrees measured by the hands anatomy and independently on the grip taxonomies. In this sense, considering the synergies occurred in some specific gestures [45] in the motion analysis assessment (Section 4.1) would allow the exoskeleton to guide the fingers with more effectiveness.

All these issues, whose resolution constitutes a natural continuation of the research activity carried out thus far, will be subjected to further investigation in the very near future.

Author Contributions: Design & Development, N.S. and M.B. (Matteo Bianchi); Supervision A.R. and Y.V.; Validation, F.V. and M.B. (Massimo Bianchini); Conceptualization, L.G. and B.A.

Funding: This work has been supported by Don Carlo Gnocchi Foundation (Italy) and by the HOLD project (Hand exoskeleton system, for rehabilitation and activities Of daily Living, specifically Designed on the patient anatomy), funded by the University of Florence.

Conflicts of Interest: The authors declare no conflict of interest.

References

1. Heller, A.; Wade, D.; Wood, V.; Sunderland, A.; Hewer, R.; Ward, E. Arm function after stroke: Measurement and recovery over the first three months. *J. Neurol. Neurosurg. Psychiatry* **1987**, *131*, 714–719. [CrossRef] [PubMed]
2. Wade, D.; Langton-Hewer, R.; Wood, V.; Skilbeck, C.; Ismail, H. The hemiplegic arm after stroke: Measurement and recovery. *J. Neurol. Neurosurg. Psychiatry* **1983**, *46*, 521–524. [CrossRef]
3. Sunderland, A.; Tinson, D.; Bradley, L.; Hewer, R. Arm function after stroke. An evaluation of grip strength as a measure of recovery and a prognostic indicator. *J. Neurol. Neurosurg. Psychiatry* **1989**, *52*, 1267–1272. [CrossRef] [PubMed]
4. Nakayama, H.; Jørgensen, H.; Raaschou, H.; Olsen, T. Recovery of upper extremity function in stroke patients: The Copenhagen Stroke Study. *Arch. Phys. Med. Rehabil.* **1994**, *75*, 394–398. [CrossRef]
5. World Health Organization. *WHO Global Disability Action Plan 2014–2021: Better Health for All People with Disability*; Technical Report; World Health Organization: Geneva, Switzerland, 2014.
6. Bruno, S.; Oussama, K. *Springer Handbook of Robotics*; Springer: Berlin, Germany, 2016.
7. Heo, P.; Gu, G.M.; Lee, S. Current hand exoskeleton technologies for rehabilitation and assistive engineering. *Int. J. Precis. Eng. Manuf.* **2012**, *13*, 807–824. [CrossRef]
8. Troncossi, M.; Mozaffari-Foumashi, M.; Parenti-Castelli, V. An Original Classification of Rehabilitation Hand Exoskeletons. *J. Robot. Mech. Eng. Res.* **2016**, *1*, 17–29. [CrossRef]
9. Bos, R.A.; Haarman, C.J.; Stortelder, T.; Nizamis, K.; Herder, J.L.; Stienen, A.H.; Plettenburg, D.H. A structured overview of trends and technologies used in dynamic hand orthoses. *J. NeuroEng. Rehabil.* **2016**, *13*, 62. [CrossRef] [PubMed]
10. Harwin, W.S.; Patton, J.L.; Edgerton, V.R. Challenges and opportunities for robot-mediated neurorehabilitation. *Proc. IEEE* **2006**, *94*, 1717–1726. [CrossRef]
11. Rosenstein, L.; Ridgel, A.L.; Thota, A.; Samame, B.; Alberts, J.L. Effects of combined robotic therapy and repetitive-task practice on upper-extremity function in a patient with chronic stroke. *Am. J. Occup. Ther.* **2008**, *1*, 28–35. [CrossRef]
12. Martinez, L.; Olaloye, O.; Talarico, M.; Shah, S.; Arends, R.; BuSha, B. A power-assisted exoskeleton optimized for pinching and grasping motions. In Proceedings of the 2010 Northeast Bioengineering Conference, New York, NY, USA, 26–28 March 2010; pp. 1–2. [CrossRef]
13. Moromugi, S.; Koujina, Y.; Ariki, S.; Okamoto, A.; Tanaka, T.; Feng, M.Q.; Ishimatsu, T. Muscle stiffness sensor to control an assistance device for the disabled. *Artif. Life Robot.* **2004**, *8*, 42–45. [CrossRef]
14. Iqbal, J.; Tsagarakis, N.; Fiorilla, A.; Caldwell, D. A portable rehabilitation device for the hand. In Proceedings of the International Conference of the IEEE Engineering in Medicine and Biology, Buenos Aires, Argentina, 1–4 September 2010; pp. 3694–3697. [CrossRef]
15. Jones, C.; Wang, F.; Osswald, C.; Kang, X.; Sarkar, N.; Kamper, D. Control and kinematic performance analysis of an Actuated Finger Exoskeleton for hand rehabilitation following stroke. In Proceedings of the International Conference on Biomedical Robotics and Biomechatronics (BioRob), Tokyo, Japan, 26–29 September 2010; pp. 282–287. [CrossRef]
16. Li, J.; Zheng, R.; Zhang, Y.; Yao, J. iHandRehab: An interactive hand exoskeleton for active and passive rehabilitation. In Proceedings of the International Conference on Rehabilitation Robotics (ICORR), Tokyo, Japan, 26–29 September 2010; pp. 1–6. [CrossRef]
17. Hasegawa, Y.; Mikami, Y.; Watanabe, K.; Sankai, Y. Five-fingered assistive hand with mechanical compliance of human finger. In Proceedings of the International Conference on Robotics and Automation (ICRA), Pasadena, CA, USA, 19–23 May 2008; pp. 718–724. [CrossRef]

18. Schabowsky, C.; Godfrey, S.; Holley, R.; Lum, P. Development and pilot testing of HEXORR: Hand EXOskeleton Rehabilitation Robot. *J. Neuroeng. Rehabil.* **2010**. [PubMed]

19. Wege, A.; Kondak, K.; Hommel, G. Mechanical design and motion control of a hand exoskeleton for rehabilitation. In Proceedings of the International Conference on Mechatronics and Automation, Niagara Falls, ON, Canada, 20 July–1 August 2005; pp. 155–519. [CrossRef]

20. Chiri, A.; Vitiello, N.; Giovacchini, F.; Roccella, S.; Vecchi, F.; Carrozza, M. Mechatronic Design and Characterization of the Index Finger Module of a Hand Exoskeleton for Post-Stroke Rehabilitation. *IEEE/ASME Trans. Mech.* **2012**, *17*, 884–894. [CrossRef]

21. Wege, A.; Hommel, G. Development and control of a hand exoskeleton for rehabilitation of hand injuries. In Proceedings of the International Conference on Intelligent Robots and Systems (IROS 2005), Edmonton, AB, Canada, 2–6 August 2005; pp. 3046–3051. [CrossRef]

22. Takahashi, C.; Der-Yeghiaian, L.; Le, V.; Motiwala, R.; Cramer, S. Robot-based hand motor therapy after stroke. *Brain* **2008**, *131*, 425–437. [PubMed]

23. Mulas, M.; Folgheraiter, M.; Gini, G. An EMG-controlled exoskeleton for hand rehabilitation. In Proceedings of the International Conference on Rehabilitation Robotics, Bellevue, WA, USA, 20–24 June 2013; pp. 371–374. [CrossRef]

24. Rotella, M.F.; Reuther, K.; Hofmann, C.; Hage, E.; BuSha, B. An orthotic hand-assistive exoskeleton for actuated pinch and grasp. In Proceedings of the Northeast Bioengineering Conference, Cambridge, MA, USA, 3–5 April 2009; pp. 1–2. [CrossRef]

25. Brokaw, E.; Black, I.; Holley, R.; Lum, P. Hand Spring Operated Movement Enhancer (HandSOME): A Portable, Passive Hand Exoskeleton for Stroke Rehabilitation. *Trans. Neural Syst. Rehabil. Eng.* **2011**, *19*, 391–399. [CrossRef] [PubMed]

26. Stilli, A.; Cremoni, A.; Bianchi, M.; Ridolfi, A.; Gerii, F.; Vannetti, F.; Wurdemann, H.A.; Allotta, B.; Althoefer, K. AirExGlove—A novel pneumatic exoskeleton glove for adaptive hand rehabilitation in post-stroke patients. In Proceedings of the 2018 IEEE International Conference on Soft Robotics, Livorno, Italy, 24–28 April 2018; pp. 579–584. [CrossRef]

27. Yap, H.; Nasrallah, F.; Lim, J.; Low, F.; Goh, J.; Yeow, R. MRC-glove: A fMRI compatible soft robotic glove for hand rehabilitation application. In Proceedings of the International Conference on Rehabilitation Robotics, Singapore, 11–14 August 2015; pp. 735–740. [CrossRef]

28. Polygerinos, P.; Wang, Z.; Galloway, K.; Wood, R.; Walsh, C.J. Soft robotic glove for combined assistance and at-home rehabilitation. *Robot. Auton. Syst.* **2015**, *73*, 135–143. [CrossRef]

29. Deimel, R.; Brock, O. A novel type of compliant and underactuated robotic hand for dexterous grasping. *Int. J. Robot. Res.* **2016**, *35*, 161–185. [CrossRef]

30. Delph, M.A.; Fischer, S.; Gauthier, P.; Luna, C.; Clancy, E.; Fischer, G. A soft robotic exomusculature glove with integrated sEMG sensing for hand rehabilitation. In Proceedings of the International Conference on Rehabilitation Robotics, Bellevue, WA, USA, 20–24 June 2013. [CrossRef]

31. Tong, K.; Ho, S.; Pang, P.; Hu, X.; Tam, W.; Fung, K.; Wei, X.; Chen, P.; Chen, M. An intention driven hand functions task training robotic system. In Proceedings of the International Conference of the IEEE Engineering in Medicine and Biology, Buenos Aires, Argentina, 31 August–4 September 2010.

32. Iqbal, J.; Khelifa, B. Stroke rehabilitation using exoskeleton-based robotic exercisers: Mini Review. *Biomed. Res.* **2015**, *26*, 197–201.

33. Lucas, L.; DiCicco, M.; Matsuoka, Y. An EMG-Controlled Hand Exoskeleton for Natural Pinching. *J. Robot. Mech.* **2004**, *16*, 1–7. [CrossRef]

34. Conti, R.; Meli, E.; Ridolfi, A.; Bianchi, M.; Governi, L.; Volpe, Y.; Allotta, B. Kinematic synthesis and testing of a new portable hand exoskeleton. *Meccanica* **2017**. [CrossRef]

35. Allotta, B.; Conti, R.; Governi, L.; Meli, E.; Ridolfi, A.; Volpe, Y. Development and experimental testing of a portable hand exoskeleton. In Proceedings of the 2015 IEEE/RSJ International Conference on Intelligent Robots and Systems, Hamburg, Germany, 28 September–2 October 2015; pp. 5339–5344. [CrossRef]

36. Byrd, R.H.; Gilbert, J.C.; Nocedal, J. A trust region method based on interior point techniques for nonlinear programming. *Math. Program.* **2000**, *89*, 149–185. [CrossRef]

37. Bianchi, M.; Fanelli, F.; Giordani, L.; Ridolfi, A.; Vannetti, F.; Allotta, B. An automatic scaling procedure for a wearable and portable hand exoskeleton. In Proceedings of the 2016 IEEE 2nd International Forum on Research and Technologies for Society and Industry Leveraging a Better Tomorrow, Bologna, Italy, 7–9 September 2016; pp. 1–5. [CrossRef]
38. Lobo-Prat, J.; Kooren, P.N.; Stienen, A.H.; Herder, J.L.; Koopman, B.F.; Veltink, P.H. Non-invasive control interfaces forintention detection in active movement-assistive devices. *J. NeuroEng. Rehabil.* **2014**, *11*, 168. [PubMed]
39. Adewuyi, A.A.; Hargrove, L.J.; Kuiken, T.A. Evaluating EMG Feature and Classifier Selection for Application to Partial-Hand Prosthesis Control. *Front. Neurorobot.* **2016**, *10*, 15. [CrossRef] [PubMed]
40. Secciani, N.; Bianchi, M.; Ridolfi, A.; Vannetti, F.; Allotta, B. Assessment of a Hand Exoskeleton Control Strategy Based on User's Intentions Classification Starting from Surface EMG Signals. In *Wearable Robotics: Challenges and Trends*; Carrozza, M.C., Micera, S., Pons, J.L., Eds.; Springer International Publishing: Cham, Switzerland, 2019; pp. 40–444.
41. Secciani, N.; Bianchi, M.; Meli, E.; Volpe, Y.; Ridolfi, A. A novel application of a surface ElectroMyoGraphy-based control strategy for a hand exoskeleton system: A single-case study. *Int. J. Adv. Robot. Syst.* **2019**, *16*. [CrossRef]
42. Bianchi, M.; Buonamici, F.; Furferi, R.; Vanni, N. Design and Optimization of a Flexion/Extension Mechanism for a Hand Exoskeleton System. In Proceedings of the ASME 2016 International Design Engineering Technical Conferences and Computers and Information in Engineering Conference, Charlotte, NC, USA, 21–24 August 2016.
43. Bianchi, M.; Cempini, M.; Conti, R.; Meli, E.; Ridolfi, A.; Vitiello, N.; Allotta, B. Design of a Series Elastic Transmission for hand exoskeletons. *Mechatronics* **2018**, *51*, 8–18. [CrossRef]
44. Santello, M.; Flanders, M.; Soechting, J. Postural hand synergies for tool use. *J. Neurosci.* **1998**, *18*, 10105–10115. [PubMed]
45. Santello, M.; Bianchi, M.; Gabiccini, M.; Ricciardi, E.; Salvietti, G.; Prattichizzo, D.; Ernst, M.; Moscatelli, A.; Jörntell, H.; Kappers, A.M.; et al. Hand synergies: Integration of robotics and neuroscience for understanding the control of biological and artificial hands. *Phys. Life Rev.* **2016**, *17*, 1–23. [CrossRef] [PubMed]

Sample Availability: Prototypes of the presented exoskeletons are available from the authors.

Article

Analysis of the Oscillating Motion of a Solid Body on Vibrating Bearers

Kuatbay Bissembayev [1,2], Assylbek Jomartov [1,*], Amandyk Tuleshov [1] and Tolegen Dikambay [2]

[1] Ministry of Education and Science of Kazakhstan, Institute Mechanics and Mechanical Engineering, Almaty 050010, Kazakhstan
[2] Ministry of Education and Science of Kazakhstan, Abai Kazakh National Pedagogical University, Almaty 050000, Kazakhstan
* Correspondence: legsert@mail.ru; Tel.: +7-777-329-5999

Received: 27 June 2019; Accepted: 3 September 2019; Published: 6 September 2019

Abstract: This article considers the oscillation of a solid body on kinematic foundations, the main elements of which are rolling bearers bounded by high-order surfaces of rotation at horizontal displacement of the foundation. Equations of motion of the vibro-protected body have been obtained. It is ascertained that the obtained equations of motion are highly nonlinear differential equations. Stationary and transitional modes of the oscillatory process of the system have been investigated. It is determined that several stationary regimes of the oscillatory process exist. Equations of motion have been investigated also by quantitative methods. In this paper the cumulative curves in the phase plane are plotted, a qualitative analysis for singular points and a study of them for stability are performed. In the Hayashi plane a cumulative curve of a body protected against vibration forms a closed path which does not tend to the stability of a singular point. This means that the vibration amplitude of a body protected against vibration does not remain constant in a steady state, but changes periodically.

Keywords: vibroprotection; seismic; rolling bearer; vibration; non-linear vibrations; cumulative curves; singular point

1. Introduction

The issue of vibration protection for devices and equipment is one of the main directions of development of the theory for vibrations of mechanical systems.

The theory of nonlinear vibration isolation has witnessed significant developments because of pressing demands for the protection of structural installations, nuclear reactors, mechanical components, and sensitive instruments from earthquake ground motion, shocks, and impact loads. In views of these demands, engineers and physicists have developed different types of nonlinear vibration isolators. This article [1] presents a comprehensive assessment of the recent developments of nonlinear isolators in the absence of active control means. It does not deal with other means of linear or nonlinear vibration absorbers. The article is closed by conclusions, which highlight resolved and unresolved problems and recommendations for future research directions.

A new, passive, vibro-protective device of the rolling-pendulum tuned mass damper type is presented [2] that, relying on a proper three-dimensional guiding surface, can simultaneously control the response of the supporting structure in two mutually orthogonal horizontal directions. Unlike existing examples of ball vibration absorbers, mounted on spherical recesses and effective for axial-symmetrical structures, the new device is bidirectional tunable, by virtue of the optimum shape of the rolling cavity, to both fundamental structural modes, even when the corresponding natural frequencies are different, in such a case recurring to an innovative non-axial-symmetrical rolling guide.

A new optimization method for a tuned mass damper (TMD) system is proposed in the paper [3], based on the artificial fish swarm algorithm (AFSA), and the primary structural damping is taken into consideration. The optimization goal is to minimize the maximum dynamic amplification factor of the primary structure under external harmonic excitations.

The paper [4] deals with the performance analysis of a vibration-isolation system for Michelangelo Buonarroti's famous Ronadanini Pietà statue based on the monitoring and analysis of vibration signals. A tuned mass-damper-inerter is introduced in order to increase the effectiveness of the isolator in horizontal direction. Specifically, a multi-degree-of-freedom (MDOF) model for the system, including non-linear terms, is proposed.

Creating vibration protection devices, using rolling bearings, is currently widespread in transport techniques to prevent transported oversize cargoes from longitudinal accelerations, for the seismic protection of structures and in other areas of modern technology. However, further progress in improving vibro-protective rolling bearings necessitates dynamic properties research and finding more advanced design solutions on the basis of this research. Most modern vibro-supporting devices use movable supports, bounded by spherical surfaces.

Articles [5,6] are focused on the technical issues, as well as on issues of improving the engineering calculation for kinematic foundations designed by this author.

Work by Y.D. Cherepinskiy [7] considers the motion of structures on the kinematic piers of a particular design, proposed by the author. It investigates a motion, originating on a plane without rolling friction, which has a significant impact on the character of the system motion.

The passive neutralization oscillations systems for high-rise construction are under consideration [8]. Their advantages and disadvantages have been revealed. A roller oscillation neutralization system for high-rise constructions subject to seismic affecting is offered. The principle of its work is described and its advantages are estimated. A mathematical movement model for carrying and carried bodies is made. Low-frequency oscillation vibration protection systems under the influence of external harmonious impact are considered. Optimum adjustment parameters for a roller damper in the structure of the compensation system are defined.

The nonlinear normal vibration modes of a mechanical system having the pendulum vibration absorber are considered [9]. The coupled and localized vibration modes are selected. In the last case the main vibration energy is concentrated in the pendulum, so this vibration mode is the most appropriate for the vibration absorption. The modes stability is investigated.

The work [10] researches low-frequency vibrations of vibro-protective system of solid bodies formed by a roller damper and a moving load-carrying body under the action of external harmonic excitations. The dynamic equations of combined motion of the working body of the damper over the hinged roller without sliding and the load-carrying body are deduced and numerically analyzed. A new procedure for evaluation of the optimal parameters of adjustment of roller dampers in nonlinear systems is proposed.

Longitudinal vibrations are investigated for the four-mass vibration-resistant system of the following solid bodies: long cargo, turnstile with roller shock absorbers, and the coupling of two flat cars after their collision with a braked hammer car [11]. The level of dynamic loads applied to the elements of the vibration-resistant system is numerically analyzed.

Low-frequency vibrations of a vibro-protection "roller damper-movable bearing body" system of rigid bodies under the action of an external harmonic excitation are considered. The working surface of the damper working body is formed by a brachistochrone. The dynamic equations of common no-slip motion of the damper working body on a hinged roller and of the bearing body are formulated. The roller damper tuning parameters are determined [12].

In all these studies, to get the final results we considered a vibration device, the bearing elements of which are bounded by spherical surfaces. A common disadvantage of all these devices is the lack of reliability at a high level of seismic disturbance. Practically, such systems behave linearly with respect to disturbance and, to suppress the vibrations of the protected bodies, such systems of seismic protection are complemented by special devices of dry friction with the specified backlash.

This work [13] studies the features of vibration motion of an orthogonal mechanism with disturbances, such as restricted power in the presence of a fixed load on the horizontal link. Dynamic and mathematical models were prepared, and the operating conditions' fields of existence for the vibration mechanism in terms of driving power were defined.

In the work [14] the mathematical expressions for the rolling resistance arising from rolling of a bearing, bounded by high order rotational surfaces are obtained.

The work [15] contains a systematic depiction of non-linear systems analysis methods, described by differential equations of second-rate. This work also contains topological and graphical methods, applicable for the calculation of autonomic and, especially, non-autonomic systems.

In the book [16] the theory of non-linear vibrations is expounded, the topic of great interest at present because of its many applications to important fields in physics and engineering.

This paper [17] presents the results of modeling of vibrations of a rigid rotor caused by the degradation of hydrodynamic bearings. The model is composed by applying equations of nonlinear hydrodynamic forces and the measured parameters of a real rotary machine.

The work contains geometrical non-linear analysis derived according to the Hamilton principle.

In the work [18] a systematic method is developed for the dynamic analysis of structures with sliding isolation, which is a highly non-linear dynamic problem. According to the proposed method, a unified motion equation can be adapted for both stick and slip modes of the system. Unlike the traditional methods by which the integration interval has to be chopped into infinitesimal pieces during the transition of sliding and non-sliding modes, the integration interval remains constant throughout the whole process of the dynamic analysis by the proposed method so that the accuracy and efficiency in the analysis of the non-linear system can be enhanced to a large extent.

The effects of neglecting small harmonic terms in the estimation of the dynamic stability of the steady state solution determined in the frequency domain are considered in the paper [19]. For that purpose, a simple single-degree-of-freedom piecewise linear system excited by a harmonic excitation is analyzed. In the time domain, steady state solutions are obtained by using the method of piecing the exact solutions (MPES) and in the frequency domain, by the incremental harmonic balance method (IHBM). The stability of the solutions obtained in the frequency domain by IHBM is determined by using the Floquet-Liapounov theorem and by digital simulation of the corresponding disturbed motion.

The aim of the present work is to study the dynamics of vibro-protection systems, the main elements of which are rolling bearings, bounded by surfaces of rotation of high order (in the absence of rolling friction).

2. Statement of the Problem

Under the influence of longstanding loads, surfaces of a rolling bearing and bases change their curvature. There are two cases: under the effect of longstanding loads, the curvature radius of the rolling supports surface and bases changes by a finite amount and forms a finite area of support. From an analytical point of view, the dynamic properties of these supports are close to the dynamic properties of the rolling support with bounded surfaces of a high-order.

Let us consider the principle of work of the kinematic foundation of moving supporting elements, which is the rolling bearing with bounded surfaces of rotation of a high (n, m) order (Figure 1).

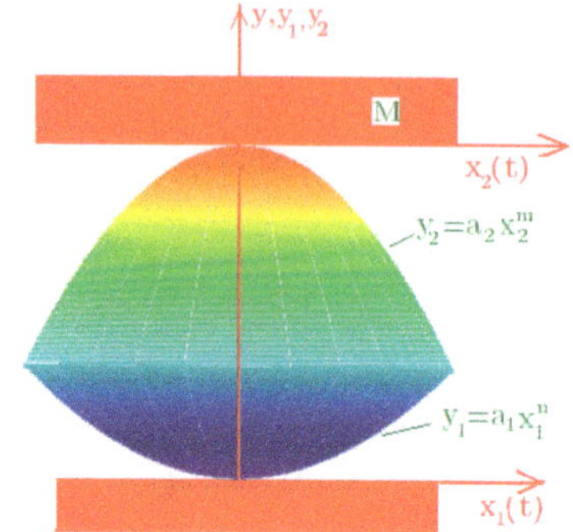

Figure 1. Scheme of rolling bearers with bearing.

Figure 1 demonstrates the rolling bearing (the object I) with bounded (top and bottom) surfaces of rotation, expressed by formulas:

$$\widetilde{y}_1 = \widetilde{a}_1 \widetilde{x}_1^{n}, \quad \widetilde{y}_2 = \widetilde{a}_2 \widetilde{x}_2^{m} \tag{1}$$

having a common axis of symmetry; but objects 2 and 3 are a stationary base (foundation) and the inner coat of a vibro-protected body. The specifics of such support is that the radius of curvature in the vicinity of the central support points tends to infinity and decreases with increasing distance from the axis of symmetry, i.e., there is straightening of the bearing surfaces in the vicinity of the central point. When considering n to infinity ($n \rightarrow \infty$), the rolling bearing I shall take a cylindrical shape.

In the systems, the restoring force arises because of the increase of potential energy when picking up the support's centre of gravity or supports and protected body. Contact with the rolling bearing surfaces of the vibro-insulated body and the foundation will be assumed as planes.

We assume that the foundation of the considered body has a small plane length, allowing us not to take into account the asynchrony of transmission of external influence from the various points of the foundation and vibro-insulated body. Equation (1) refers to the coordinate system associated with the rolling bearings (see Figure 2). The curvature radius of the vertices of these surfaces at $n, m > 2$ tends to infinity, i.e., there is straightening of the bearing surfaces. Let us denote the horizontal offset of the bases as $\widetilde{x}_0(t)$. As $\widetilde{x}(t)$ we denote a displacement of the upper body, supporting on the rolling bearing.

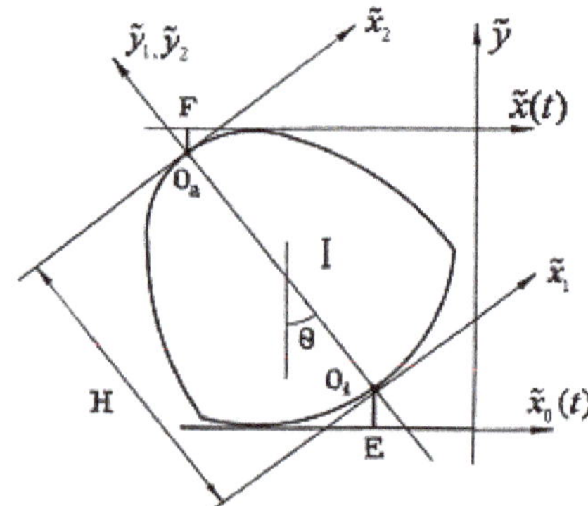

Figure 2. Scheme of coordinates of rolling bearing surfaces of high order.

In Figure 2 the rolling bearing is shown in the position when the base and body are offset relative to each other by $(\widetilde{x} - \widetilde{x}_0)$.

We define the dependence between the horizontal relative displacement of the foundation of a resilient construction on rolling bearers and their vertical shift. Let us introduce a new notation (Figure 2):

$$\theta = \frac{\widetilde{x} - \widetilde{x}_0}{H}$$

where H and θ are the height and angle of the rotational displacement, respectively.

On the other hand

$$\theta = \frac{d\widetilde{y}_1}{d\widetilde{x}_1} = \frac{d\widetilde{y}_2}{d\widetilde{x}_2}, \quad \theta = n\widetilde{a}_1\widetilde{x}_1^{n-1} = m\widetilde{a}_2\widetilde{x}_2^{m-1} \tag{2}$$

Through Equation (2), we express $\widetilde{x}_1$, $\widetilde{y}_1$ and $\widetilde{x}_2$, $\widetilde{y}_2$ via θ as

$$\widetilde{x}_1 = \frac{\theta^{\frac{1}{n-1}}}{(n\widetilde{a}_1)^{\frac{1}{n-1}}}, \quad \widetilde{y}_1 = \frac{\widetilde{a}_1}{(n\widetilde{a}_1)^{\frac{1}{n-1}}}\theta^{\frac{n}{n-1}}, \quad \widetilde{x}_2 = \frac{\theta^{\frac{1}{m-1}}}{(m\widetilde{a}_2)^{\frac{1}{m-1}}}, \quad \widetilde{y}_2 = \frac{\widetilde{a}_2}{(m\widetilde{a}_2)^{\frac{1}{m-1}}}\theta^{\frac{m}{m-1}} \tag{3}$$

Vertical displacement of the foundation is written as:

$$\widetilde{y} = H\cos\theta = H + O_1E + O_2F, \quad \widetilde{y} = -2H\sin^2\frac{\theta}{2} = H + O_1E + O_2F \tag{4}$$

where

$$O_1E = \widetilde{x}_1\sin\theta - \widetilde{y}_1\cos\theta, \quad O_2F = \widetilde{x}_2\sin\theta - \widetilde{y}_2\cos\theta$$

Transforming the function $\widetilde{y}$ to the range of Taylor and taking into account the first order of smallness of the angle θ, for relation Equation (4) we get the expression:

$$\widetilde{y} = -H\frac{\theta^2}{2} + \widetilde{x}_1\theta - \widetilde{y}_1 + \widetilde{x}_2\theta - \widetilde{y}_2 \tag{5}$$

Substituting expression Equations (3)–(5), we get

$$\widetilde{y} = -H\frac{\theta^2}{2} + \frac{(n-1)\widetilde{a}_1}{(n\widetilde{a}_1)^{\frac{n}{n-1}}}\theta^{\frac{n}{n-1}} + \frac{(m-1)\widetilde{a}_2}{(m\widetilde{a}_1)^{\frac{m}{m-1}}}\theta^{\frac{m}{m-1}} \tag{6}$$

Taking into account the expression Equation (2), we rewrite the expression Equation (6) in the form

$$\widetilde{y} = -\frac{1}{2H}(\widetilde{x} - \widetilde{x}_0)^2 + \frac{(n-1)\widetilde{a}_1}{(Hn\widetilde{a}_1)^{\frac{n}{n-1}}}(\widetilde{x} - \widetilde{x}_0)^{\frac{n}{n-1}} + \frac{(m-1)\widetilde{a}_2}{(Hm\widetilde{a}_1)^{\frac{m}{m-1}}}(\widetilde{x} - \widetilde{x}_0)^{\frac{m}{m-1}} \tag{7}$$

Term Equation (7) defines the dependence between the horizontal relative movements of the bodies' bases on the rolling bearings with straightened surfaces and their vertical displacements. In the case, when $n = m$, term Equation (7) takes the form

$$\widetilde{y} = -\frac{1}{2H}(\widetilde{x} - \widetilde{x}_0)^2 + \frac{(n-1)}{(Hn)^{\frac{n}{n-1}}}\left(\frac{1}{\sqrt[n-1]{\widetilde{a}_1}} + \frac{1}{\sqrt[n-1]{\widetilde{a}_2}}\right)(\widetilde{x} - \widetilde{x}_0)^{\frac{n}{n-1}} \tag{8}$$

3. The Equation of Motion of the Vibro-Protected Bodies on the Rolling Bearings with Straightened Surfaces

To derive the differential equations of motion of a body, we use the equations of Ferrers, considering the Equation (8) as a holonomic link, superimposed on the vertical movement of the body. Kinetic and potential energy of the vibro-protected bodies are expressed as

$$T = M\frac{\dot{\widetilde{x}}^2 + \dot{\widetilde{y}}^2}{2}, \quad U = Mg\widetilde{y} \tag{9}$$

where $g-$ free-fall acceleration. The equation of motion of the body on the rolling bearings, in case of small oscillations, have the form:

$$\ddot{\widetilde{x}} + \frac{g}{H}\left[\frac{1}{(H\widetilde{a}_1 n)^{\frac{1}{n-1}}(\widetilde{x}-\widetilde{x}_0)^{\frac{n-2}{n-1}}} + \frac{1}{(H\widetilde{a}_2 m)^{\frac{1}{m-1}}(\widetilde{x}-\widetilde{x}_0)^{\frac{m-2}{m-1}}} - 1\right](\widetilde{x}-\widetilde{x}_0) = 0 \tag{10}$$

So, movement of the protected body (even in a case of small oscillations) is to be described by a nonlinear equation. Let us consider the case, when $n = m$, then Equation (10) will have the form

$$\ddot{\widetilde{x}} + \omega_0^2\left[\frac{\widetilde{N}_n}{\sqrt[n-1]{(\widetilde{x}-\widetilde{x}_0)^{n-2}}} - 1\right](\widetilde{x}-\widetilde{x}_0) = 0 \tag{11}$$

where

$$\widetilde{N}_n = \frac{1}{\sqrt[n-1]{nH}}\left[\frac{1}{\sqrt[n-1]{\widetilde{a}_1}} + \frac{1}{\sqrt[n-1]{\widetilde{a}_2}}\right], \quad \omega_0^2 = \frac{g}{H} \tag{12}$$

At $n \to \infty$, the Equation (11) becomes nonlinear equation of the form

$$\ddot{x}(t) - \omega_0^2 x(t) + 2\omega_0^2 \mathrm{sign}x(t) = -\ddot{x}_0(t), \tag{13}$$

describing oscillations of a body on the supports, having a rectangular shape.

The nonlinear Equation (6) describes the motion of the vibro-protected bodies on the rolling bearings, bounded by a parabola of higher order.

Let us introduce a new notation:

$$x = \frac{\widetilde{x}}{H}, \quad x_0 = \frac{\widetilde{x}_0}{H}, \quad t = \omega_0 \tau \tag{14}$$

The Equation (11) can be reduced to an equation in dimensionless form

$$\ddot{x} + \Phi(x - x_0) - x = -x_0(t) \tag{15}$$

where

$$\Phi(x - x_0) = N_n(x - x_0)^{\frac{1}{n-1}} \tag{16}$$

$$\widetilde{N}_n = \frac{1}{\sqrt[n-1]{nH}}\left[\frac{1}{\sqrt[n]{\widetilde{a}_1}} + \frac{1}{\sqrt[n]{\widetilde{a}_2}}\right], \quad a_1 = \widetilde{a}_1 H^{n-1}, \quad a_2 = \widetilde{a}_2 H^{n-1} \tag{17}$$

4. The Study of Free Oscillations of a Body on the Rolling Bearings with Straightened Surfaces

At $x_0 = 0$ the Equations (6) and (9) take the form

$$\ddot{x} + \Phi(x) - x = 0 \tag{18}$$

As per the method of restructuring [4], we represent the solution and the nonlinear term of Equation (10) as a truncated trigonometric series

$$x = \sum_{k=1}^{\nu} A_{2k-1}\sin(2k-1)\psi, \quad \Phi(x) = \sum_{k=1}^{\nu} b_{2k-1}\sin(2k-1)\psi \tag{19}$$

where $\psi = \omega(A_1)t..$

By choosing the required number of collocation points in interval $0 \leq \psi \leq 2\pi$, we obtain the system of algebraic equations

$$
\left.
\begin{aligned}
a_{11}b_{11} + a_{13}b_{13} + \ldots + a_{1(2v-1)}b_{1(2v-1)} &= \Phi\!\left(\sum_{k=1}^{v} A_{2k-1}a_{1(2k-1)}\right) = \Phi_1(A_1, A_3, \ldots, A_{2v-1}) \\
a_{21}b_{21} + a_{23}b_{23} + \ldots + a_{2(2v-1)}b_{2(2v-1)} &= \Phi\!\left(\sum_{k=1}^{v} A_{2k-1}a_{2(2k-1)}\right) = \Phi_2(A_1, A_3, \ldots, A_{2v-1}) \\
a_{v1}b_{v1} + a_{v3}b_{v3} + \ldots + a_{v(2v-1)}b_{v(2v-1)} &= \Phi\!\left(\sum_{k=1}^{v} A_{2k-1}a_{v(2k-1)}\right) = \Phi_v(A_1, A_3, \ldots, A_{2v-1})
\end{aligned}
\right\}
\tag{20}
$$

where

$$
a_{i(2k-1)} = \sin(2k-1)\psi_i, \, i = 1, 2, \ldots v; k = 1, 2, \ldots v,
$$

ψ_i is the argument term at the collocation point i. Determinator of system Equation (20) is not equal to zero for arbitrarily chosen collocation points inside the period.

Permitting this system Equation (20) in relation to coefficients b_{2k-1}, we obtain

$$
b_{2k-1} = \sum_{i=1}^{v} a_{i(2k-1)}^{(-1)} \Phi_i(A_1, A_3, \ldots, A_{2v-1})
\tag{21}
$$

Now by substituting expression Equation (19) in Equation (18) and equating the coefficients at the similar harmonics $\sin(2k-1)\psi$, we obtain v equations

$$
-(2k-1)^2\omega^2 A_{2k-1} + b_{2k-1}(A_1, A_3, \ldots, A_{2v-1}) = 0, \, k = 1, 2, \ldots, v
\tag{22}
$$

Regard these equations as

$$
\omega^2 = \frac{b_1(A_1, A_3, \ldots, A_{2v-1})}{A_1}, \quad A_{2k-1} = \frac{b_k(A_1, A_3, \ldots, A_{2v-1})}{(2k-1)^2\omega^2}, \, (k = 2, 3, \ldots, v).
\tag{23}
$$

Thus, the first equation expresses a value of the frequency point of self-oscillations of the non-linear system through amplitudes of harmonic solution. The following equation determines amplitudes of higher harmonics.

Now the Equation (23) is suitable for implementation of the iterative method. Solutions, received by the iteration method, in many cases come together, since expressions for amplitudes $2k - 1$ for harmonics A_{2k-1} are inversely proportional to the multiplier $(2k-1)^2\omega^2$.

The method of iteration is convenient to use by specifying value

$$
A_1 \neq 0; A_3 = A_5 = \ldots = A_{2v-1} = 0,
$$

The first Equation (23) takes the form of amplitude-frequency characteristics, and the remaining equations allow the definition of the form of self-oscillations of the system, presented by unabridged trigonometric range.

Assigning ψ value to $\frac{\pi}{6}, \frac{\pi}{4}, \frac{\pi}{2}$ and being limited to the terms $k = 1, 2, 3$ in (21), we obtain the system as

$$
b_1 = \frac{N_n}{3}\left[\left(\tfrac{1}{2}A_1 + A_3 + \tfrac{1}{2}A_5\right)^{\frac{1}{n-1}} + \sqrt{3}\left(\tfrac{\sqrt{3}}{2}A_1 - \tfrac{\sqrt{3}}{2}A_5\right)^{\frac{1}{n-1}} + (A_1 - A_3 + A_5)^{\frac{1}{n-1}}\right],
$$

$$
b_3 = \frac{N_n}{3}\left[2\left(\tfrac{1}{2}A_1 + A_3 + \tfrac{1}{2}A_5\right)^{\frac{1}{n-1}} - (A_1 - A_3 + A_5)^{\frac{1}{n-1}}\right],
$$

$$
b_5 = \frac{N_n}{3}\left[\left(\tfrac{1}{2}A_1 + A_3 + \tfrac{1}{2}A_5\right)^{\frac{1}{n-1}} - \sqrt{3}\left(\tfrac{\sqrt{3}}{2}A_1 - \tfrac{\sqrt{3}}{2}A_5\right)^{\frac{1}{n-1}} + (A_1 - A_3 + A_5)^{\frac{1}{n-1}}\right].
$$

For the first approximation, let us assume that $A_1 \neq 0, A_3 = 0, A_5 = 0$, we obtain

$$b_1 = N_n K_1 A_1^{\frac{1}{n-1}}, b_3 = N_n K_3 A_1^{\frac{1}{n-1}}, b_5 = N_n K_5 A_1^{\frac{1}{n-1}}, a = 1, \tag{24}$$

where

$$K_1 = \tfrac{1}{3}\left[\frac{1}{2^{\frac{1}{n-1}}} + \sqrt{3}\left(\frac{\sqrt{3}}{2}\right)^{\frac{1}{n-1}} + 1\right] K_3 = \tfrac{1}{3}\left[2^{\frac{n-2}{n-1}} - 1\right], K_5 = \tfrac{1}{3}\left[\frac{1}{2^{\frac{1}{n-1}}} - \sqrt{3}\left(\frac{\sqrt{3}}{2}\right)^{\frac{1}{n-1}} + 1\right]$$

$$\varphi = \omega(A_1) \cdot (t + t_0).$$

Substituting Equation (24) in expression Equation (23) we define that

$$\omega^{(1)} = \sqrt{\frac{N_n K_1}{A_1^{\frac{n-2}{n-1}}} - 1}, \; A_3 = \frac{N_n K_1}{9\left(\omega^{(1)}\right)^2 + 1} A_1^{\frac{1}{n-1}}, \; A_5 = \frac{N_n K_5}{25\left(\omega^{(1)}\right)^2 + 1} A_1^{\frac{1}{n-1}}, \tag{25}$$

The amount in Formula (19) is chosen by the odd harmonics for reasons of the symmetry of the oscillating system. When holding in the sum Equation (19) the even terms (in the process of constructing a solution), the coefficients of these terms become zero.

To determine the second approximation we believe that $A_1 \neq 0, A_3 \neq 0, A_5 = 0$.

Solving the equations of motion Equation (18) in the second approximation has the form:

$$x = A_1 \sin \omega^{(2)} t + A_3^{(2)} \sin 3\omega^{(2)} t + A_2^{(2)} \sin 5\omega^{(2)} t,$$

where

$$\omega^{(2)} = \sqrt{\left(\frac{N_n K_1}{A_1^{\frac{n-2}{n-1}}} - 1\right) - \widetilde{\Omega}_n}, \; A_3^{(2)} = \frac{N_n \widetilde{K}_3 A_1^{\frac{1}{n-1}}}{9\left(\omega^{(2)}\right)^2 + 1}, \; A_5^{(2)} = \frac{N_n \widetilde{K}_5 A_1^{\frac{1}{n-1}}}{25\left(\omega^{(2)}\right)^2 + 1},$$

$$\widetilde{\Omega}_n = \frac{N_n A_1^{\frac{1}{n-1}}}{3}\left[\left(\frac{1}{2^{\frac{1}{n-1}}} + 1\right) - \left(\tfrac{1}{2} + \frac{N_n K_3 A_1^{-\frac{n-2}{n-1}}}{9\left(\omega^{(1)}\right)^2 + 1}\right)^{\frac{1}{n-1}} - \left(1 - \frac{N_n K_3 A^{-\frac{n-2}{n-1}}}{9\left(\omega^{(1)}\right)^2 + 1}\right)^{\frac{1}{n-1}}\right],$$

$$\widetilde{K}_5 = \tfrac{1}{3}\left[\left(\tfrac{1}{2} + \frac{N_n K_3 A_1^{-\frac{n-2}{n-1}}}{9\left(\omega^{(1)}\right)^2 + 1}\right)^{\frac{1}{n-1}} - \sqrt{3}\left(\frac{\sqrt{3}}{2}\right)^{\frac{1}{n-1}} + \left(1 - \frac{N_n K_3 A_1^{-\frac{n-2}{n-1}}}{9\left(\omega^{(1)}\right)^2 + 1}\right)^{\frac{1}{n-1}}\right], \tag{26}$$

$$\widetilde{K}_3 = \tfrac{1}{3}\left[2\left(\tfrac{1}{2} + \frac{N_n K_3 A_1^{-\frac{n-2}{n-1}}}{9\left(\omega^{(1)}\right)^2 + 1}\right)^{\frac{1}{n-1}} - \left(1 - \frac{N_n K_3 A_1^{-\frac{n-2}{n-1}}}{9\left(\omega^{(1)}\right)^2 + 1}\right)^{\frac{1}{n-1}}\right],$$

As an example, we consider the oscillations of a vibro-protected body on the rolling bearings, the supporting surfaces of which are bounded by the parabolas of the fourth and sixth degree, with the following parameter values,

$$n = 4; \; \widetilde{a}_1 = 6,25 \cdot 10^{-8} \, sm^{-3}; \; \widetilde{a}_2 = 15 \cdot 10^{-8} \, sm^{-3}; n = 6; \; \widetilde{a}_1 = 1,56 \cdot 10^{-12} \, sm^{-5}; \; \widetilde{a}_2 = 6,6 \cdot 10^{-12} \, sm^{-5};$$
$$H = 3m; \; \omega_0^2 = 3,26 s^{-2}; \; g = 9,8 m/s^2.$$

Dependence of the system frequency on the amplitude built by the trigonometric collocation method is shown in Figure 3. The dotted line, carried out in the graph, is built on the basis of the second approximation Equation (26). The solid line is the curve of the first approximation Equation (25).

The proximity of the curves gives an idea of the rate of convergence of the iterative processes. Considering the natural vibration frequencies to infinity, at the amplitude $A_1 \to 0$ for nonlinear systems, there is a «clash». Thus, for vibro-protection systems, a bearing element of which is a high-order parabola, the «clash» phenomenon is typical for small oscillations.

The oscillation frequency of the system slowly decreases with increasing amplitude. A parabola of the second order (for small oscillations) is independent of the amplitude (Figure 3).

Figure 4 shows the graphs of the solutions, obtained by analytical methods (line 1) and by quantitative integration by using the Runge-Kutta scheme (line 2).

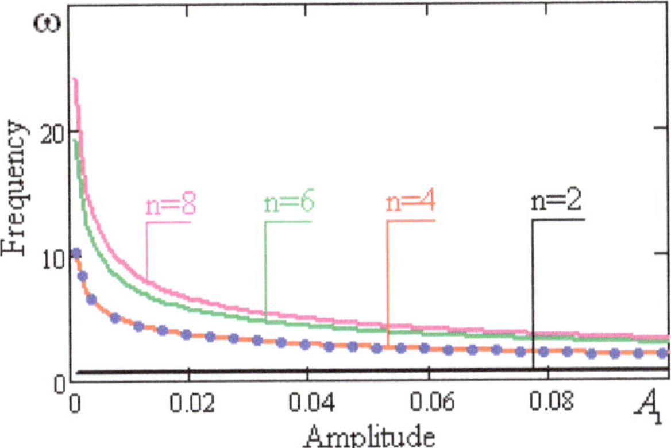

Figure 3. Graphs of dependence of self-oscillation oscillation frequency on various means of *n* equation.

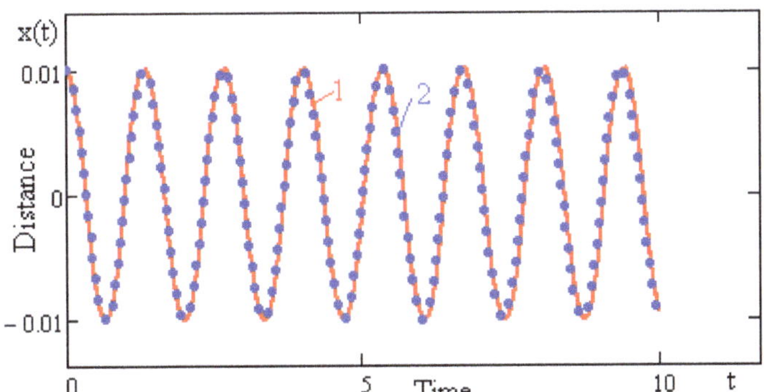

Figure 4. Graphs of solutions of free frequency on various means of *n* equation.

A comparison of lines 1 and 2 shows very good exactness of the analytical calculation.

5. Forced Oscillations of a Vibro-Protected Body, Caused by a Movable Base

Let us study the vibrations of a body at harmonic horizontal displacement of the lower base

$$x_0(t) = Q \sin pt, \tag{27}$$

where Q and p-dimensionless amplitude and frequency of disturbance s.

Write the solution and the nonlinear term of Equation (15) in the form

$$x = \sum_{k=1}^{v} A_{2k-1} \sin(2k-1)pt, \Phi(x-x_0) = \sum_{k=1}^{v} b_{2k-1} \sin(2k-1)pt. \tag{28}$$

Substituting Equation (28) to the equation of motion Equation (15) and limited by $k = 1, 2, 3$, we obtain a system of equations

$$-\left(p^2+1\right)A_1 + b_1\left(A_1, A_3, A_5\right) = -Q, -\left(9p^2+1\right)A_3 + b_3(A_1, A_3, A_5) = 0,$$
$$-(25p^2+1)A_5 + b_5(A_1, A_3 A_5) = 0. \tag{29}$$

To apply the method of iteration, we present this system of equations in the form

$$A_1 = \frac{Q}{p^2+1} + \frac{b_1(A_1,A_3,A_5)}{p^2+1}, \quad A_3 = \frac{b_3(A_1,A_3,A_5)}{9p^2+1}, \quad A_5 = \frac{b_5(A_1,A_3,A_5)}{25p^2+1}. \tag{30}$$

The coefficients of the trigonometric series Equation (28) b_{2k-1} are determined by the method of collocation. Identifying expressions Equations (16) and (28), as well as considering the relation Equations (28) and (27), we get the equation in the form

$$\sum_{k=1}^{\nu} b_{2k-1}\sin(2k-1)\varphi = N_n\Big(\sum_{k=1}^{\nu} A_{2k-1}\sin(2k-1)\varphi - Q\sin\varphi\Big)^{\frac{1}{n-1}}, \tag{31}$$

where $\varphi = pt$.

Giving to φ values $\pi/6, \pi/3, \pi/2$, we get a system of equations relatively b_1, b_2, b_3 :

$$b_1 - b_3 + b_5 = N_n[(A_1-Q)-A_3+A_5]^{\frac{1}{n-1}}, \quad \tfrac{\sqrt{3}}{2}b_1 - \tfrac{\sqrt{3}}{2}b_5 = N_n\Big[\tfrac{\sqrt{3}}{2}(A_1-Q)-\tfrac{\sqrt{3}}{2}A_5\Big]^{\frac{1}{n-1}},$$
$$\tfrac{1}{2}b_1 + b_3 + \tfrac{1}{2}b_5 = N_n\Big[\tfrac{1}{2}(A_1-Q)+A_3+\tfrac{1}{2}A_5\Big]^{\frac{1}{n-1}}, \tag{32}$$

from which we can find

$$b_1 = \frac{N_n}{3}\Bigg[\frac{1}{\sqrt{3}}\Big(\tfrac{1}{2}(A_1-Q)+A_3+\tfrac{1}{2}A_5\Big)^{\frac{1}{n-1}} + \Big(\tfrac{\sqrt{3}}{2}(A_1-Q)-\tfrac{\sqrt{3}}{2}A_5\Big)^{\frac{1}{n-1}} +$$
$$+ \tfrac{1}{\sqrt{3}}\big((A_1-Q)-A_3+A_5\big)^{\frac{1}{n-1}}\Bigg],$$
$$b_3 = \frac{N_n}{3}\Bigg[2\Big(\tfrac{1}{2}(A_1-Q)+A_3+\tfrac{1}{2}A_5\Big)^{\frac{1}{n-1}} - \big((A_1-Q)-A_3+A_5\big)^{\frac{1}{n-1}}\Bigg], \tag{33}$$
$$b_5 = \frac{N_n}{3}\Bigg[\Big(\tfrac{1}{2}(A_1-Q)+A_3+\tfrac{1}{2}A_5\Big)^{\frac{1}{n-1}} - \sqrt{3}\Big(\tfrac{\sqrt{3}}{2}(A_1-Q)-\tfrac{\sqrt{3}}{2}A_5\Big)^{\frac{1}{n-1}} +$$
$$+ \big((A_1-Q)-A_3+A_5\big)^{\frac{1}{n-1}}\Bigg].$$

Assuming in Equation (33) that $A_1 \neq 0, A_3 = A_5 = 0$, we obtain the values of the coefficients in the first approximation

$$b_1 = N_n K_1(A_1-Q)^{\frac{1}{n-1}}, \quad b_3 = N_n K_3(A_1-Q)^{\frac{1}{n-1}}, \quad b_5 = N_n K_5(A_1-Q)^{\frac{1}{n-1}}, \tag{34}$$

Taking into account Equation (34), we rewrite the relation Equation (30) in the form

$$A_1^{(1)} = \frac{1}{p^2+1}\Big[Q + N_n K_1\big(A_1^{(1)}-Q\big)^{\frac{1}{n-1}}\Big], \quad A_3^{(1)} = \frac{N_n K_3}{9p^2+1}\big(A_1^{(1)}-Q\big)^{\frac{1}{n-1}}, \quad A_5^{(1)} = \frac{N_n K_5}{25p^2+1}\big(A_1^{(1)}-Q\big)^{\frac{1}{n-1}} \tag{35}$$

Substituting Equation (35) in (28), we get the solution of Equation (15)

$$x = C_1 \sin pt + \frac{N_n K_3}{9p^2+1}(A_1-Q)^{\frac{1}{n-1}}\sin 3pt + \frac{N_n K_5}{25p^2+1}(A_1-Q)^{\frac{1}{n-1}}\sin 5pt. \tag{36}$$

6. Results and Analysis

As an example for the parameters, given in Section 4. we have built a graph, determining the dependence of the system amplitude on the frequency of disturbances corresponding to first, third, and fifth harmonics at $Q = 3 \times 10^{-3}$ value of kinematic disturbances amplitude (See Figures 5 and 6).

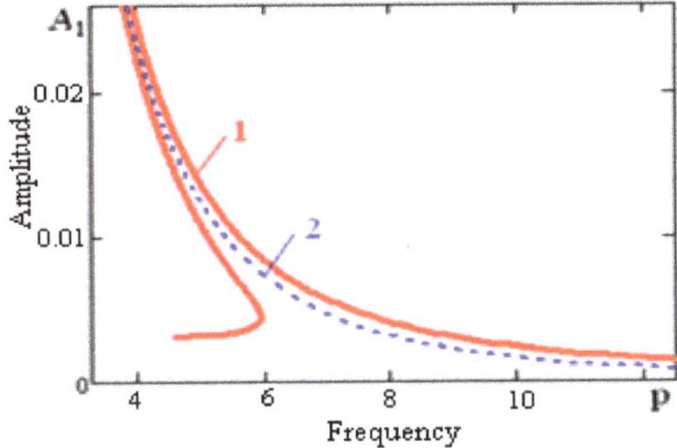

Figure 5. Amplitude-frequency response curve for fundamental harmonic.

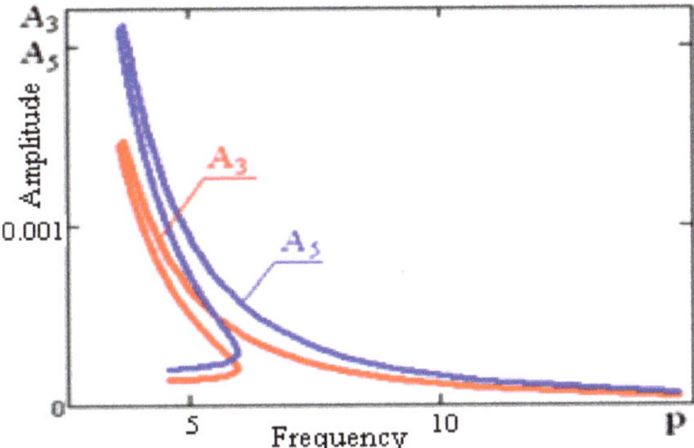

Figure 6. Amplitude-frequency response curve or third and fifth harmonics.

In Figure 5, line 1 corresponds to the resonance line, line 2 to the structural. In Figure 7 the resonance lines are shown corresponding to various means n.

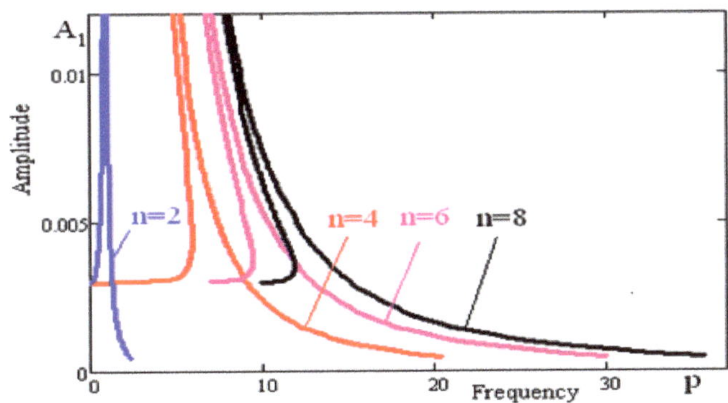

Figure 7. Graph of amplitude-frequency features of the main harmonic for various means of n.

Thus, the analysis of Figure 5 allows us to draw the following conclusions:

I. Vibratory bearings on rolling nodes, bounded by surfaces of rotation of a high order, can be attributed to nonlinear systems with soft characteristics.

II. The amplitude of the forced oscillations (up to the area of the resonant frequency) maintains a constant value, and in the resonant state, amplitude decreases to zero.

III. The calculation shows that the amplitudes of the harmonics of a higher order are smaller, as compared with the amplitudes of the fundamental harmonics (Figure 6). This demonstrates the closeness of the oscillatory process to the harmonic.

Inertia force, acting on the vibro-protected objects, is weakly dependent on the amplitude of disturbance (Figure 8). When changing the amplitude of the kinematic disturbance eight times (6–9 points), the inertia force increases in the range of 30% of the initial value.

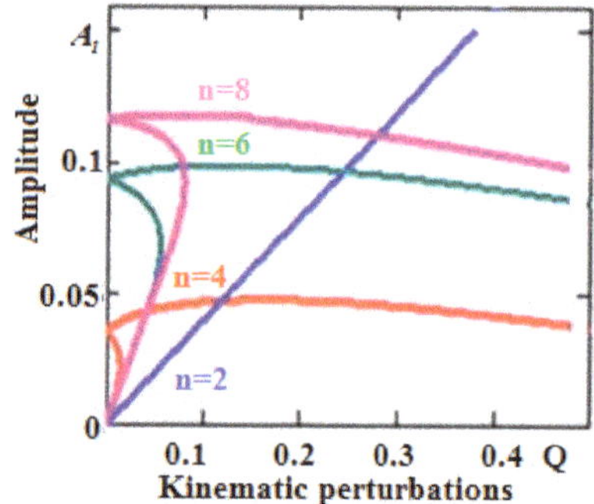

Figure 8. Graph of dependence of amplitude of vibro-protected body on amplitude of kinematic disturbance for various means of n.

For comparison we indicate that for the spherical bearings, inertia force acting on the vibro-protected body grows proportional to the amplitude of the kinematic disturbance, typical for all the linear systems. This property of the rolling bearings, bounded by the surfaces of the high order, makes them promising for creating vibration protective structures under strong kinematic disturbance.

In Figures 9 and 10, two-dimensional projections of a phase-portrait and dispensation of spectral concentration (Fourier-spectrum) of periodic oscillations of the vibro-protected bodies on the rolling bearings with straightened surfaces at $p = 5.65$ value of kinematic disturbance frequency are shown.

It is of interest to note that the range of many-fold harmonics to kinematic disturbance frequency appears in the spectrum of responses of vibro-protected systems.

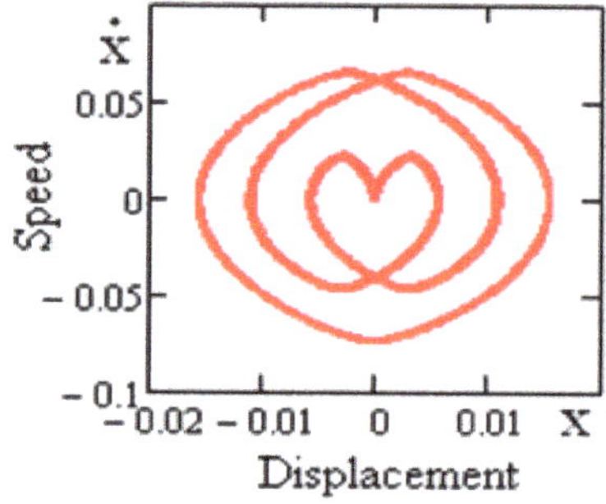

Figure 9. Trajectory of vibro-protected body on phase plain.

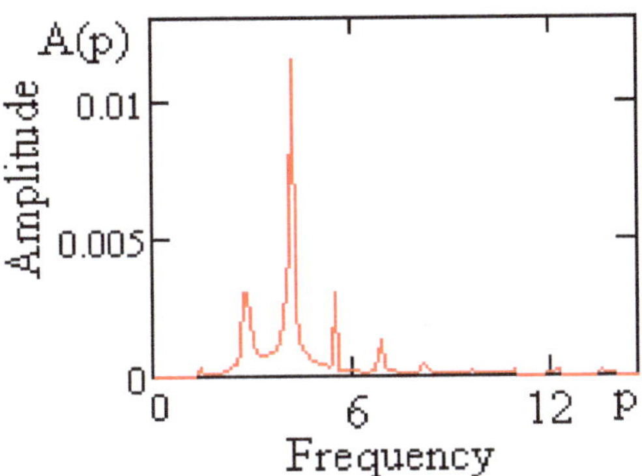

Figure 10. Allocation of spectral density of periodic oscillations of vibro-protected body.

7. Stability of Periodic Solutions

Assuming that in the case of harmonic oscillations, a component of the fundamental frequency, having the period $2\pi/p$, dominates over the higher harmonics. The periodic solution and first derivative of Equation (15) can be approximately represented as follows:

$$x = a \sin pt + b \cos pt, \; \dot{x} = ap \cos pt - bp \sin pt, \tag{37}$$

Let us suppose that the amplitudes a and b are functions of time and slowly vary depending on t. For the nonlinear term of Equation (15), Fourier series expansion looks as:

$$\Phi(x - x_0) = N_n C^{\frac{1}{n-1}} \sin^{\frac{1}{n-1}}(pt + \gamma) = \sum_{k=1}^{\infty} B_{2k-1} \sin(2k-1)pt + D_{2k-1} \cos(2k-1)pt \tag{38}$$

where

$$C = \sqrt{(a-Q)^2 + b^2}, \; tg\gamma = \frac{b}{a-Q}, \; B_{2k-1} = N_n K_{2k-1} \frac{(a-Q)}{\left[(a-Q)^2 + b^2\right]^{\frac{n-2}{2(n-1)}}},$$

$$D_{2k-1} = N_n K_{2k-1} \frac{b}{\left[(a-Q)^2 + b^2\right]^{\frac{n-2}{2(n-1)}}}, \; K_{2k-1} = \sqrt{L_{2k-1}^2 + M_{2k-1}^2}, \tag{39}$$

$$L_{2k-1} = \frac{1}{\pi} \int_0^{2\pi} \sin^{\frac{1}{n-1}}\psi \sin(2k-1)\psi d\psi, \; M_{2k-1} = \frac{1}{\pi} \int_0^{2\pi} \sin^{\frac{1}{n-1}}\psi \cos(2k-1)\psi d\psi, \; \psi = pt + \gamma.$$

Substituting Equations (37) and (38) in (15) and equating to zero the individual coefficients of the terms, containing $\sin pt$ and $\cos pt$, we have

$$\frac{da}{dt} = \frac{1}{p} \left\{ (p^2 + 1) - N_n K_1 \frac{1}{\left[(a-Q)^2 + b^2\right]^{\frac{n-2}{2(n-1)}}} \right\} b = X(a,b),$$

$$\frac{db}{dt} = -\frac{1}{p} \left\{ \left[(p^2 + 1) - N_n K_1 \frac{1}{\left[(a-Q)^2 + b^2\right]^{\frac{n-2}{2(n-1)}}} \right] (a-Q) + p^2 Q \right\} = Y(a,b). \tag{40}$$

Let us consider the steady state, when amplitudes $a(t)$ and $b(t)$ in (37) are constant, i.e.,

$$\frac{da}{dt} = X(a,b) = 0, \frac{db}{dt} = Y(a,b) = 0. \tag{41}$$

In light of these conditions, from Equations (40) we can obtain that the set amplitude $a_0 = A$, $b_0 = 0$ of the periodic solution $x(t)$ is determined by the formula

$$A = \frac{1}{p^2+1}\left[N_n K_1 (A-Q)^{\frac{1}{n-1}} + Q\right]. \tag{42}$$

Let us derive the conditions for the stability of periodic solutions. We will consider small deviations ξ and η from the amplitudes a_0 and b_0, and will find out when these deviations (with increasing time) are close to zero.

From Equation (40) we get

$$\frac{d\xi}{dt} = \alpha_1 \xi + \alpha_2 \eta, \frac{d\eta}{dt} = \beta_1 \xi + \beta_2 \eta, \tag{43}$$

where

$$\alpha_1 = \frac{(n-2)}{(n-1)}\frac{1}{p}\frac{W_0}{C_0^2}(a_0 - Q)b_0, \alpha_2 = \frac{1}{p}\left\{(p^2+1) - W_0 + \frac{(n-2)}{(n-1)}\frac{W_0}{C_0^2}b_0^2\right\},$$
$$\beta_1 = \frac{1}{p}\left\{-(p^2+1) + W_0 - \left(\frac{n-2}{n-1}\right)\frac{W_0}{C_0^2}(a_0 - Q)^2\right\}, \beta_2 = -\frac{1}{p}\left\{\left(\frac{n-2}{n-1}\right)\frac{W_0}{C_0^2}(a_0 - Q)b_0\right\}, \tag{44}$$

where

$$W_0 = \frac{N_n K_1}{C_0^{\frac{n-2}{n-1}}}, C_0 = A - Q.$$

The characteristic equation of the system has the form:

$$\lambda^2 - (\alpha_1 + \beta_2)\lambda + \alpha_1\beta_2 - \alpha_2\beta_1 = 0.a \tag{45}$$

The stability condition is given by the Routh-Hurwitz criteria, i.e.,

$$\alpha_1 + \beta_2 = 0, (\alpha_1 = 0, \beta_2 = 0).\alpha_1\beta_2 - \alpha_2\beta_1 > 0$$

to

$$\left[(p^2+1) - W_0\right]\left[(p^2+1) - \frac{W_0}{n-1}\right] > 0. \tag{46}$$

The singular point, i.e., steady system state, is a center.

The boundary of unstable periodic solutions of Equation (40) is determined by the curves.

$$p^2 = W_0 - 1, p^2 = \frac{W_0}{n-1} - 1, \tag{47}$$

and stability areas are determined by the following inequalities [15,16]

$$p^2 - (W_0 - 1) > 0, p^2 - \left(\frac{W_0}{n-1} - 1\right) > 0, p^2 - (W_0 - 1) < 0, p^2 - \left(\frac{W_0}{n-1} - 1\right) < 0. \tag{48}$$

In Figure 11 resonant curves are drawn by Equation (42). There are two branches in these graphs with respective positive and negative means of amplitude A. In Figure 11 borders of stability, built by Formula (47) are shown by lines 3; 4 and 5; 6 (lines 3; 4 for positive, lines 5; 6 for negative amplitude). From the physical perspective, positiveness of amplitude signifies phase coincidence, but negativeness of amplitudes signifies opposite-phase. Tangent lines to resonant lines are parallel to axes 7.

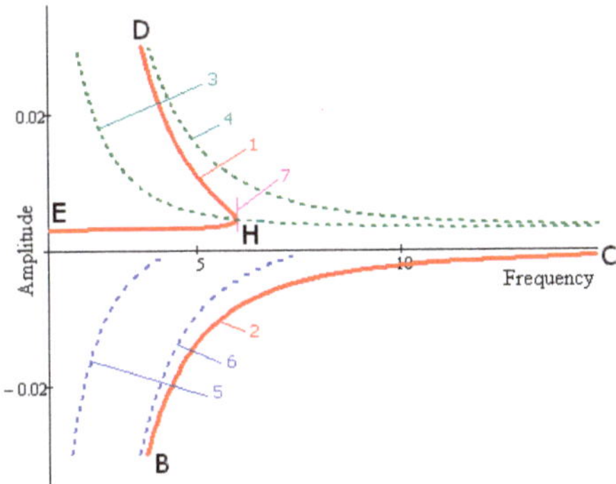

Figure 11. Graph of amplitude-frequency characteristics and of lines of borders of stability field.

Point H divides the upper part of the graph into stable HD and instable parts EH. The lower part of the graph is stable. Calculations were made in the following values of parameters:

$$n = 4; \ \widetilde{a}_1 = 6,25 \cdot 10^{-8} sm^{-3}; \ \widetilde{a}_2 = 15 \cdot 10^{-8} sm^{-3}; H = 3m;$$

8. Numerical Studies

The idea of seismic isolation of buildings in case of earthquakes with the help of rolling element linear guides is one of the simplest and most efficient ideas in the history of earthquake-proof construction [20].

The operation principle of earthquake protection devices based on rolling element linear guides is that it provides for the mobility of the building footing, which decreases the inertial force acting on it.

The seismic impacts that occur during an earthquake are classified as random actions, although in simplified calculation models that are used in practice, they are usually treated as determinate. The magnitude and nature of seismic impacts cannot be accurately predicted in advance. The instrumental records that characterize the change patterns of seismic impacts, which are characteristic of individual earthquakes, over time, never repeat one another even if they occur in the same place. Therefore, we can only discuss their similarity and, therefore, their classification in general terms.

The work [21] contain earthquake scales with their basic characteristics are listed in Table 1.

Table 1. Main characteristics of an earthquake.

Intensity (Points)	Maximum Displacement Intervals, *sm*	Maximum Speed Intervals, $\frac{sm}{n}$	Maximum Acceleration Intervals, $\frac{sm}{n^2}$
6	0.15–0.3	3–6	30–60
7	0.31–0.6	6.1–12	61–120
8	0.61–0.12	12.1–24	121–240
9	0.121–0.24	24.1–48	241–480

At a seismic level no more than 6 points, no special protection measures against earthquakes are taken. The work [21] shows that for short-period seismic disturbances ($T \leq 0.6c$), maximum acceleration values occur in ground motion with frequencies $p = 15\frac{1}{c} \div 20\frac{1}{c}$, and for long-wave disturbances-with basic frequencies $p = 6\frac{1}{c} \div 10\frac{1}{c}$. The original accelerograms provided in the paper show that the maximum values of seismic accelerations, short-period and long-period, are related to the frequencies $p = 10\frac{1}{c}$ and $20\frac{1}{c}$. Each given implementation of the ground motion process during the earthquake can be expanded into a Fourier series, taking the duration of the earthquake as the period. If we assume that the structure responds to one of the harmonic curves in this expansion that has a frequency closest to the natural frequency of the structure, then, neglecting short-term transient processes at the beginning and the end of the earthquake, we can treat the process of the structure motion as stationary for a limited time. When seismic waves pass, both horizontal and vertical motions of the structure footing will take place. The vertical component of the acceleration usually has a somewhat smaller amplitude than the horizontal component (some 30–50%). In calculations of seismic effects, the displacement of the footing can be taken as the sum of sinusoids. To simplify the calculation, it is usually assumed that the horizontal displacement of the base is determined by the following relationship $x_0 = Q \sin pt$.

The equations of motion of a vibration-protected body on rolling element linear guides limited by surfaces of high-order rotation Equation (11) are solved numerically using Mathcad-15 software package by the Runge-Kutta method with variable step under zero initial conditions. For the numerical analysis, six rolling element linear guide models have been selected and divided into three groups. Table 2 shows the parameters of the selected rolling element linear guide models. The coefficients of the rolling element linear guide surface are calculated using the formula $y = ax^n = \frac{1}{2R^{n-1}}x^n$, where R is the radius of the surface of the second order ($n = 2$). $\widetilde{a}_1$ and $\widetilde{a}_2$ are the coefficients of the lower and upper surfaces of the rolling element linear guide, respectively, H—Height, n—The order of the rolling element linear guide surface. The parameter $\widetilde{N}_n$ is calculated using the Formula (12). In the first option, different surface coefficients $\widetilde{a}_1$ and $\widetilde{a}_2$ of the rolling element linear guide were selected with constant n and H. In the second option, different heights H of the rolling element linear guide were selected with constant, $\widetilde{a}_1\,\widetilde{a}_2$, and n. In the first option, different orders of surfaces n of the rolling element linear guide were selected with constant $\widetilde{a}_1, \widetilde{a}_2$, and H.

Table 2. Parameters of rolling element linear guides.

Option	Number of Rolling Element Linear Guide Model	R_1 *sm*	R_2 *sm*	$\widetilde{a}_1$ *sm*$^{-(n-1)}$	$\widetilde{a}_2$ *sm*$^{-(n-1)}$	n	H *sm*	$\widetilde{N}_n$ *sm*$^{\frac{n-2}{n-1}}$
	1	200	150	6.25×10^{-8}	1.481×10^{-7}	4	300	41.497
Option 1	2	200	100	6.25×10^{-8}	5×10^{-7}	4	300	35.569
	3	150	50	1.481×10^{-7}	4×10^{-6}	4	300	23.713
	1	200	150	6.25×10^{-8}	1.481×10^{-7}	4	300	41.497
Option 2	2	200	150	6.25×10^{-8}	1.481×10^{-7}	4	200	47.502
	3	200	150	6.25×10^{-8}	1.481×10^{-7}	4	100	59.849
	1	200	150	6.25×10^{-8}	1.481×10^{-7}	4	300	41.497
Option 3	2	200	150	6.25×10^{-8}	1.481×10^{-7}	6	300	89.788
	3	200	150	3.91×10^{-17}	1.481×10^{-7}	8	300	127.112

To demonstrate the effectiveness of vibration isolation properties of the bearings confined by surfaces of rotation of higher order is test table, which shows the dependence of maximum acceleration vibroseismic body on rolling support on the intensity of the earthquake at various values of coefficients of surface bearings. According to the test table, selecting the parameters of the bearings can achieve a reduction of the acceleration vibroseismic body on rolling support.

Figure 12 shows the displacement, speed, and acceleration graphs of the vibration-protected body on rolling element linear guides in forced oscillation mode. In this case, the parameters of the rolling element linear guide surface in the simulation are chosen for the first model from the first option. The frequency and amplitude of the disturbance are $p = 17.4\frac{1}{c}, Q = 0.9\ cm$.

The main characteristics of the motion of a vibration-protected body on rolling element linear guides are as follows: maximum displacements x_{max}, maximum speed $\dot{x}_{max}$, and maximum accelerations $\ddot{x}_{max}$. Figure 13 shows the dependences of $x_{max}, \dot{x}_{max}$ on the disturbance frequency for the first model of the rolling element linear guide option. The disturbance amplitude is $Q = 0.9\ cm$. The analysis of Figure 13 makes it possible to make the following conclusion: vibration mounts on rolling bearing units limited by rotation surfaces of high-order can be regarded as non-linear systems with soft characteristics. The maximum displacements and speeds of a vibration-protected body on rolling element linear guides have two resonant frequencies. The maximum displacement values increase slowly up to the section of resonant frequencies and decrease to zero in the resonant state.

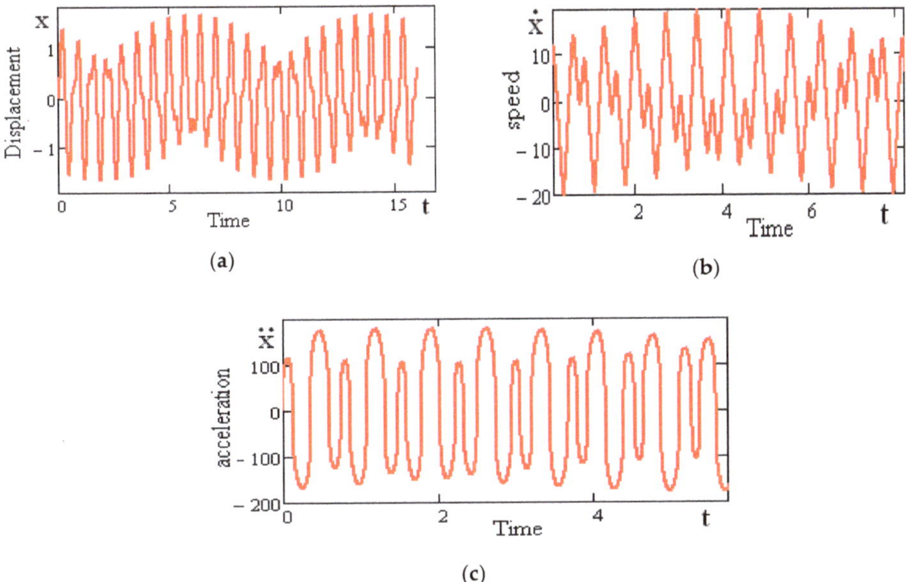

Figure 12. Dependence graph of: (**a**) displacement, (**b**) speed, and (**c**) acceleration of the vibration-protected body on rolling element linear guides on time.

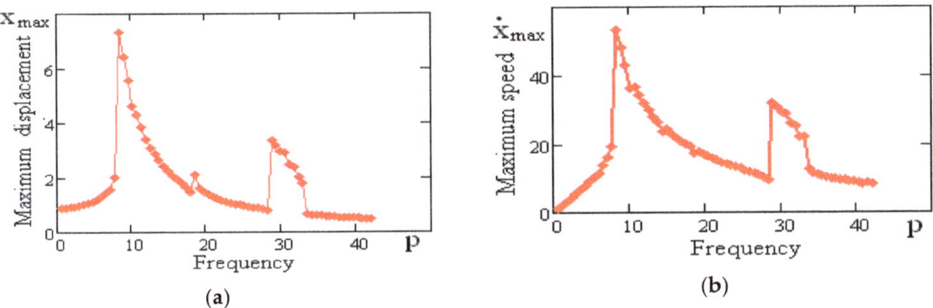

Figure 13. Dependence graph of maximum value of displacement (**a**) and speed (**b**) of the vibration-protected body on the rolling element linear guides on the frequency of the kinematic disturbance.

The calculation of the dynamic impacts on structures is of great importance in the design of structures. It allows determining the true load-bearing capacity of the structure more correctly. Assuming a vertical static load per rack at $10^4 kg$, calculations of the reaction forces are carried out.

Figure 14 shows the dependence of the maximum reaction force of a vibration-protected body on rolling element linear guides on the disturbance amplitude for the first rolling element linear guide model: curve-1 has been constructed using the numerical method, and curve-2-using the analytical method. Similar curves are shown in Figure 13, which gives an idea of how close the results of analytical and numerical calculations are. The frequency of the kinematic disturbances is $p = 17\frac{1}{c}$.

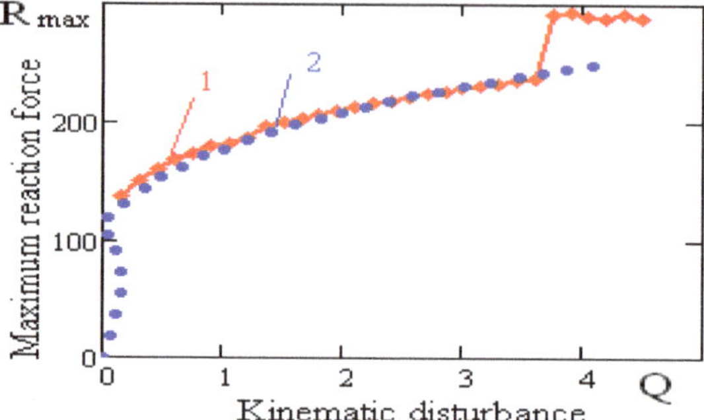

Figure 14. Dependence graph of maximum reaction force of the vibration-protected body on rolling element linear guides on the amplitude of the kinematic disturbance.

Figure 15 shows the dependence of the maximum reaction force of the vibration-protected body on rolling element linear guides on the frequency and amplitude of the disturbance for different values of the surface coefficients provided a constant surface order and rolling element linear guides height values (first option of Table 2). The figure shows that the maximum values of the reaction force of the vibration-protected body on rolling element linear guides decrease as the value of the rolling element linear guides surface coefficient increases. The maximum values of the reaction force of a vibration-protected body are weakly dependent on the amplitude of the kinematic disturbance. Resonance frequencies do not change significantly.

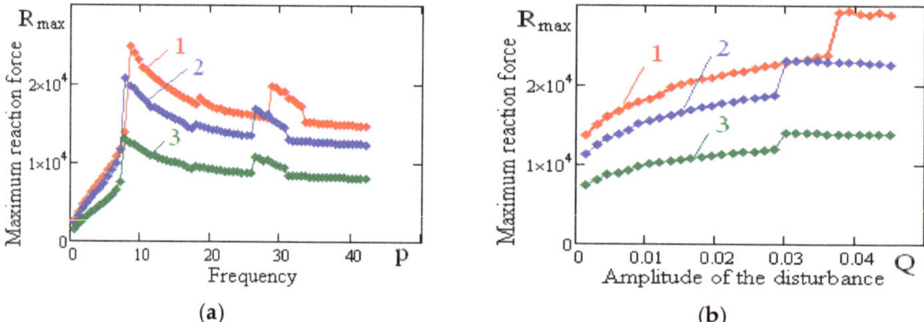

Figure 15. Dependence graph of the maximum reaction force of the vibration-protected body on rolling element linear guides on the frequency (**a**) and amplitude (**b**) of the disturbance for different values of the coefficients of the rolling element linear guide surface.

Figure 16 shows the dependence of the maximum reaction force of the vibration-protected body on rolling element linear guides on the frequency and amplitude of the disturbance for different height values provided constant values of the coefficients and the rolling element linear guide surface order (second option of Table 2). From the figure, we can see that the maximum values of the reaction force of the vibration-protected body on rolling element linear guides increase as the value of rolling element linear guide height decreases. Abrupt changes are observed in the dependency of the maximum value of the reaction force on the amplitude of the kinematic disturbance. After the jump, the maximum values of the reaction force slowly change with the increasing value of kinematic disturbance amplitude. The resonance frequencies shift toward increasing frequency of the kinematic disturbance.

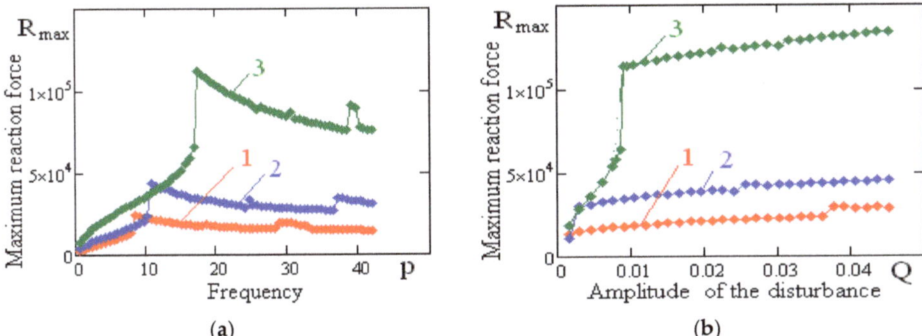

Figure 16. Dependence graph of the maximum reaction force of the vibration-protected body on rolling element linear guides on the frequency (**a**) and amplitude (**b**) of the disturbance for different values of rolling element linear guide height.

Figure 17 shows the dependence of the maximum reaction force of the vibration-protected body on rolling element linear guides on the frequency and amplitude of the disturbance for various values of the surface order provided constant values of the surface coefficients and the rolling element linear guide height (third option of Table 2). The maximum values of the reaction force of the vibration-protected body on rolling element linear guides increase as the order of the vibration-protected body on rolling element linear guide surface increases. The maximum values of the reaction force slowly change as the value of the amplitude of the kinematic disturbance increases. The resonance frequencies shift toward the increasing frequency of the kinematic disturbance.

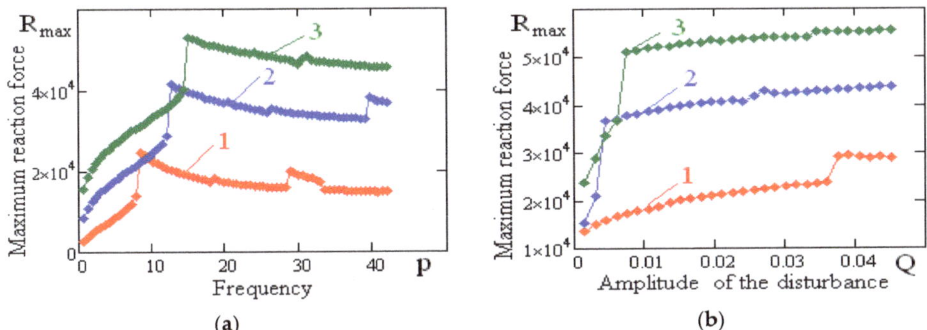

Figure 17. Dependence graph of the maximum reaction force of the vibration-protected body on rolling element linear guides on the frequency (**a**) and amplitude (**b**) of the disturbance for different values of surface order of the rolling element linear guide.

Thus, the analysis of the calculation allows us to draw the following conclusions: rolling element linear guides limited by high-order rotation surfaces can be regarded as non-linear systems with soft characteristics. The maximum values of the reaction force acting on the vibration-protected objects are weakly dependent on the amplitude of the kinematic disturbance. For comparison, we note that, for spherical bearings, the inertial forces acting on the vibration-protected bodies increase in proportion to the amplitude of the kinematic disturbance, which is characteristic of all linear systems.

This property of rolling element linear guides limited by high-order rotation surfaces makes them a promising solution for creating means of vibration protection for structures in the conditions of strong kinematic excitations.

9. Conclusions

A new mathematical model has been built and the dynamic features of vibro-protected devices, the main elements of which are rolling bearings, bounded by the surfaces of rotation of high order, have been investigated. It has been ascertained that such vibro-protected devices are highly nonlinear and a clash phenomenon appears for these systems.

In the spectrum of response emerges the range of harmonics, multiple to the frequency of kinematic disturbances. It is determined that inertial force, working on vibro-protected body on such bearings, depends little on the kinematic disturbance level.

This feature of rolling bearings, bounded by surfaces of high order, gives them potential for creation of the devices for vibro-protection of buildings under conditions of strong kinematic disturbances.

Author Contributions: Conceptualization, K.B. formal analysis, A.T.; investigation, K.B., A.J.; software, T.D.

Funding: This research was funded by Ministry of education and science of Kazakhstan grant number AP05134148.

Conflicts of Interest: The authors declare no conflict of interest.

References

1.   Ibrahim, R.A. Recent advances in nonlinear passive vibration isolators. *J. Sound Vib.* **2008**, *314*, 371–452. [CrossRef]
2.   Matta, E.; De Stefano, A.; Spencer, B.F. A new passive rolling-pendulum vibration absorber using a non-axial-symmetrical guide to achieve bidirectional tuning. *Earthq. Eng. Struct. Dyn.* **2009**, *38*, 1729–1750. [CrossRef]
3.   Shi, W.; Wang, L.; Lu, Z.; Zhang, Q. Application of an artificial fish swarm algorithm in an optimum tuned mass damper design for a pedestrian bridge. *Appl. Sci.* **2018**, *8*, 175. [CrossRef]
4.   Siami, A.; Cigada, A.; Karimi, H.R.; Zappa, E. Vibration Protection of a Famous Statue against Ambient and Earthquake Excitation Using A Tuned Inerter–Damper. *Machines* **2017**, *5*, 33. [CrossRef]
5.   Smirnov, V.Y. Testing of buildings with seismic isolation systems the dynamic loads and the real earthquakes, Earthquake-resistant construction. *Saf. Struct.* **2009**, *4*, 16–22.
6.   Erzhanov, S.E.; Lapin, V.A. The system of seismic isolation of buildings and structures in the Republic of Kazakhstan and developed countries. *Study Seism. Stab. Build. Struct.* **2015**, *23*, 164–192.
7.   Cherepinskiy, D. The seismic isolation of residential buildings. *Res. Seism. Stab. Build. Struct.* **2015**, *23*, 438–462.
8.   Burtseva, O.A.; Tkachev, A.N.; Chipko, S.A. Roller seismic impact oscillation neutralization system for high-rise buildings. *Proc. Eng.* **2015**, *129*, 259–2655. [CrossRef]
9.   Klimenko, A.A.; Mikhlin, Y.V. Nonlinear normal vibration of a mechanical system having the pendulum vibration absorber. *Mech. Solids* **2010**, *40*, 162–171.
10. Legeza, V.P. Dynamics of vibroprotective systems with roller dampers of low-frequency vibrations. *Strength Mater.* **2004**, *36*, 185–194. [CrossRef]
11. Parcianello, E.; Chisari, C.; Amadio, C. Optimal design of nonlinear viscous dampers for frame structures. *Soil Dyn. Earthq. Eng.* **2017**, *100*, 257–260. [CrossRef]
12. Legeza, V.P. Application of the theory of roller shock absorbers to the vibroprotection of transport structures. *Strength Mater.* **2006**, *38*, 214–219. [CrossRef]
13. Legeza, V.P. Efficiency of a vibro protection system with an isochronous roller damper. *Mech. Solids* **2013**, *48*, 168–177. [CrossRef]
14. Bissembayev, K.; Iskakov, Z. Oscillations of the orthogonal mechanism with a non-ideal source of energy in the presence of a load on the operating link. *Mech. Mach. Theory* **2015**, *92*, 153–170. [CrossRef]
15. Bissembayev, K.; Omyrzhanova, O.; Sultanova, K. Oscillations specific for the homogeneous rod like elastic structure on the kinematic absorber basis with rolling bearers having straightened surfaces. *Mech. Mach. Sci.* **2019**, *68*, 187–195.
16. Hayashi, T. *Non-Linear Oscillations in Physical Systems*; Princeton University Press: Princeton, NJ, USA, 2014; p. 406.
17. Schmidt, G.; Tondl, A. *Non-Linear Vibrations*; Cambridge University Press: Cambridge, UK, 2009; p. 424.

18. Jonušas, R.; Juzėnas, E.; Juzėnas, K.; Meslinas, N. Modelling of rotor dynamics caused by of degrading bearings. *Mechanics* **2012**, *18*, 438–441. [CrossRef]
19. Wang, Y.P.; Chung, L.L.; Liao, W.H. Seismic response analysis of bridges isolated with friction pendulum bearings. *Earthq. Eng. Struct. Dyn.* **1998**, *27*, 1069–1093. [CrossRef]
20. Wolf, H.; Stegi, M. The influence of neglecting small harmonic terms on estimation of dynamical stability of the response of non-linear oscillators. *Comput. Mech.* **1999**, *24*, 230–237. [CrossRef]
21. Shebalin, N.V.; Aptikaev, F.F. The Development of MSK-64 Scales. *Comput. Seismolog.* **2003**, *34*, 210–253.

Article

Nonlinear Response of Tilting Pad Journal Bearings to Harmonic Excitation [†]

Enrico Ciulli * and Paola Forte *

Department of Civil and Industrial Engineering, University of Pisa, Largo Lazzarino, 56122 Pisa, Italy
* Correspondence: enrico.ciulli@unipi.it (E.C.); paola.forte@unipi.it (P.F.)
† This paper is an extended version of the paper published in Ciulli, E.; Forte, P. Nonlinear Effects in the Dynamic Characterization of Tilting Pad Journal Bearings. In the proceedings of the International Conference of IFToMM ITALY, Cassino, Italy, 29–30 November 2018.

Received: 19 April 2019; Accepted: 14 June 2019; Published: 17 June 2019

Abstract: In the experimental identification of dynamic bearing coefficients, usually small perturbations around the static equilibrium position are assumed and linear coefficients are considered. In the literature, studies on non-linear effects in plain journal bearings, especially from a numerical point of view, are reported. Few similar studies can be found on tilting pad journal bearings (TPJB). The present work reports some peculiar aspects observed during the experimental identification procedure of TPJB linear dynamic coefficients. The tests are performed on a test bench designed for large size journal bearings operating at high peripheral speeds and static loads. A quasi-static procedure is developed to quickly check the results obtained from the usually adopted dynamic excitation. It consists of applying a slowly rotating force to the floating stator and measuring the relative displacement of the stator from the rotating shaft. Different levels of static and dynamic load are applied to two different TPJBs with four and five pads. Deformed orbits have been observed increasing the ratio between dynamic load and static load, suggesting the presence of non-linearity. Similar results are obtained with simple analytical models assuming suitably tuned non-linear stiffness terms.

Keywords: tilting pad journal bearing; nonlinear behavior; experimental characterization

---

1. Introduction

Tilting pad journal bearings (TPJB) are widely used in turbomachinery because of their stability characteristics at high speeds. For design purposes it is essential to know the coefficients that characterize their dynamic behavior. These coefficients are usually obtained with the assumption of small amplitude motions with respect to the static equilibrium position, which allows linearization according to the well-known Lund model [1] and many different experimental procedures have been proposed to identify them [2,3].

For plain journal bearings, Qiu [4] has experimentally verified that the error in the estimation of dynamic coefficients not considering nonlinearity is negligible when the perturbation displacement amplitude is in the order of magnitude of 0.02 of radial clearance (c) and suggests displacement amplitudes not exceeding 0.05c to contain the error within 2.5%.

However, in industrial practice, the amplitude of the shaft vibrating motion can reach about 0.1c in operating conditions, violating the hypothesis of small perturbations. Moreover, in experimental tests for the identification of linearized dynamic coefficients, the choice of imposing small perturbations conflicts with the need for measuring small displacements with reduced errors.

From an analytical/numerical point of view, several authors have tackled the problem of nonlinearity, especially in plain journal bearings. The approach consists of the direct integration of the Reynolds equation [5] and the description of the oil-film forces with a larger number of dynamic coefficients with respect to the classical linearized eight ones, by retaining more terms of the Taylor series expansion [6,7]. The proposed nonlinear models proved to represent well the nonlinear effects within the considered journal bearings. A parametric analysis to study the sensitivity of non-linear forces in journal bearings of different types is reported in Reference [8]. Results are presented for a two-lobe elliptical bearing showing the effects of several parameters on size, shape, and orientation of the elliptical orbits produced by synchronous vibrations.

As far as the TPJB is concerned, the literature offers less contributions compared to plain journal bearings. From the analytical/numerical point of view, nonlinear effects were studied by direct integration of the Reynolds equation in References [9–14], showing how the rotor orbit increasingly deviates from the theoretical elliptical orbit of the linear case as the ratio of dynamic and static load amplitudes increases. The effect of different unbalanced loads on a four-pad TPJB in the load-between-pad (LBP) configuration was investigated considering thermal and deformation effects in Reference [9]. Slightly triangular journal orbits were found with a dynamic load about 70% of the static one while nearly quadrilateral orbits were found with a dynamic load greater than the static one. The position and the shape of the orbits appeared to be influenced by thermal effects on both viscosity and pad deformations, while the influence of the elastic deformations appeared to be smaller. Thermo-elasto-hydrodynamic theory should be used to predict more accurate results as shown also in Reference [10]. The effect of the liner compliance on the nonlinear dynamic behavior of the same TPJB used in References [9,10] supporting vertical and horizontal rotors was investigated in Reference [11]. Almost square orbits were found when the unbalanced load was applied to the vertical rotor (with zero static load) whilst three-lobed orbits were found with a dynamic load 50% of the static one applied to the horizontal rotor. Particularly for the horizontal configuration, shape and size of the orbits are influenced by several factors such as liner thickness and material, preload, pivot offset, and radial clearance. Three-lobed orbits were also found for a TPJB with two loaded pads in LBP configuration [12,13] and five-lobed orbits for a five-pad bearing [14] also with an approximated analytical solution using Fourier series developments. Other researchers have proposed models with an increased number of coefficients identifying them from the dynamic response obtained by direct integration [15].

From the experimental point of view different TPJBs were tested in Reference [16] with two different test rigs detecting typical non-linear behaviors. Symptoms of a non-linear behavior of the pad journal bearings of two industrial machines were found in Reference [17]. The dynamic coefficients of three nonlinear models were identified with a different number of coefficients (28, 24, 36) on a five-pad TPJB bearing, with a 100 mm diameter [18]. The results of the three nonlinear models and the linear one are in good agreement regarding the identified linear terms. Moreover, the identified stiffness and damping coefficients appear to decrease with increasing dynamic force amplitude with a reduction of up to 60% for direct stiffness in the case of a dynamic force increase from 5% to 30% of the static load.

Following a preliminary investigation on possible nonlinear effects in a large size TPJB [19], the present paper focuses on the nonlinear response of tilting pad journal bearings to harmonic excitation observed during the experimental procedure for the identification of the linear dynamic coefficients. In particular it uses the tests to ascertain the linear and nonlinear range of displacements prior to the dynamic characterization campaign. Nonlinear effects related to the dynamic/static load ratio should in fact be considered in the experimental identification procedure and accounted for or avoided. There are very few experimental results published on this topic and in particular on large size TPJB. De Falco et al. [16] and Chatterton et al. [18] have dealt with the problem with smaller bearings and different test rig configurations. Moreover, as far as the authors are aware, the application of an asynchronous rotating force instead of harmonic forces with constant directions, with different dynamic/static load ratios, is new. The tests are performed on a unique experimental apparatus realized for large size journal bearings operating at high peripheral speeds and static loads, with single tone or multi-tone dynamic loads. The design criteria are described in Reference [20]. The main characteristics of the realized bench and the first experimental results obtained during the commissioning with a four-pad TPJB are reported in [21,22]. Results obtained under stationary and slow variable conditions are particularly reported in Reference [21]. The different systems related to the test rig are shown in more detail in Reference [22] together with the procedures adopted for both static and dynamic tests.

2. Materials and Methods

Figure 1 shows a picture and a schematic drawing of the test bench. The rotor, supported by rolling bearings, is driven by an electric motor connected to a gearbox with a gear ratio of six. A torque meter measures the driving torque. The test bearing housing is floating, and the static load and the dynamic ones are applied to it by three hydraulic actuators. The static load acts upwards in the vertical direction while the dynamic loads are applied in mutual orthogonal directions, at 45° with respect to the vertical one (Figure 2). The dynamic actuators can work one at a time or simultaneously. In the second case, if they operate with equal amplitude in phase, they can produce a vertical force (y direction), in antiphase a horizontal force (x direction), and in quadrature a rotating force. Three pitch stabilizers, placed at 120° around the bearing, provide the bearing housing with axial constraints that can be adjusted to align the bearing with respect to the rotor. Table 1 summarizes the main characteristics of the test rig.

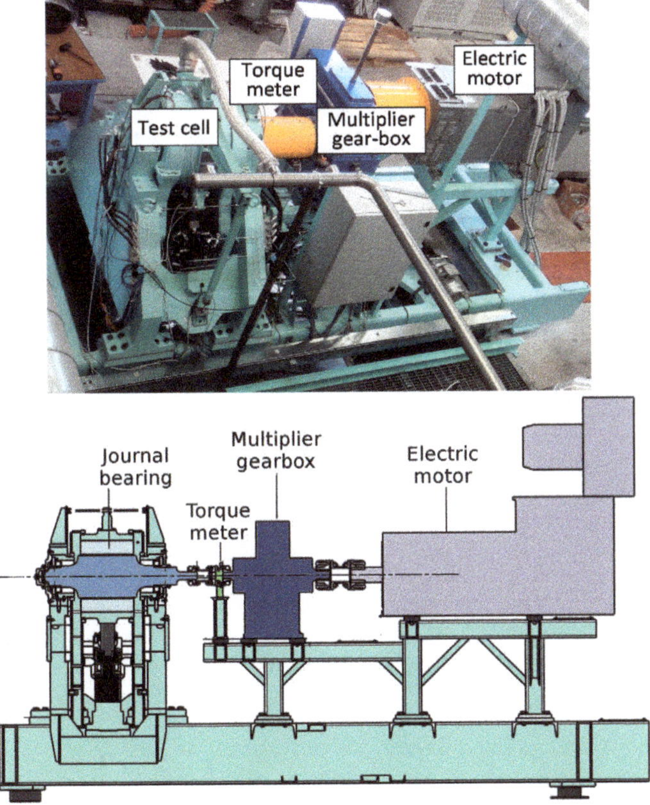

Figure 1. Photograph and schematic drawing of the test bench.

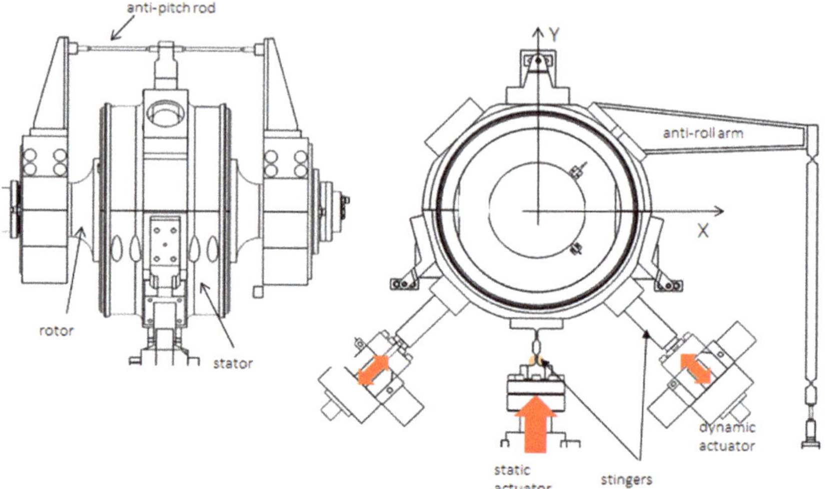

Figure 2. Drawing of the test cell.

Table 1. Main characteristics of the experimental apparatus.

Characteristic	Value Range
Bearing diameter [mm]	150–300
Bearing length to diameter ratio	0.4–1
Shaft rotational speed [rpm]	0–24,000
Bearing peripheral speed [m/s]	0–150
Static load [kN]	0–270
Dynamic load [kN]	0–40
Dynamic load frequency [Hz]	0–350
Bearing oil flow rate [L/min]	125–1100
Bearing oil inlet temperature [°C]	30–120
Electric motor power [kW]	630
Plant maximum total power [kW]	1000

Load cells and instrumented stingers, capable of measuring dynamic loads, measure all significant forces acting on the bearing housing while high-resolution proximity sensors measure the relative displacements of the bearing housing and the rotor in the directions of the dynamic actuators shown in Figure 2 (U and V directions shown in Figure 3). Eight sensors are employed, placed on two parallel planes perpendicular to the bearing axis. Four accelerometers measure the acceleration of the stator at the mid-section in the direction of the dynamic actuators.

Three different oil supply systems are used for the TPJB, the actuators, and the multiplication gearbox.

Tests are managed by a very complex control and data acquisition system. Up to 30 high-frequency signals (forces and torque, displacements, rotational speed, accelerations) can be acquired and sampled at up to 100 kHz, while up to 60 low-frequency (quasi-static) signals (temperatures, pressures, and flow-rates in the main and auxiliary lubrication systems and motor electric current) can be acquired and sampled at 1 Hz. A high sampling rate means an accurate description of signals but also a large amount of stored data; however, since the identification tests are rather short, data storage is not a problem. High-frequency signals are also averaged every second and their mean values are stored together with the low-frequency ones.

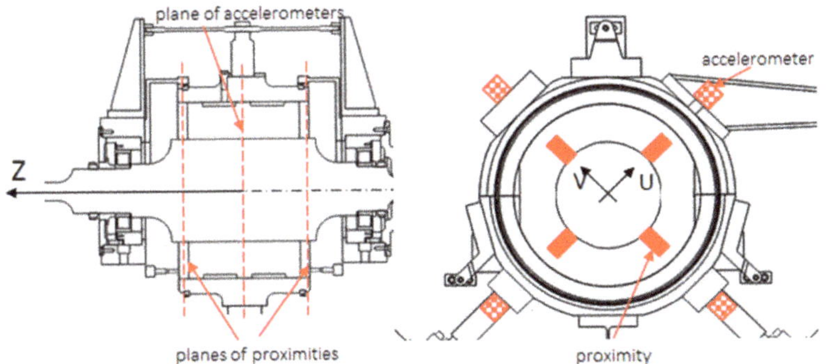

Figure 3. Test cell displacement and acceleration sensor positions.

The test articles consisted of a four-pad (without offset) and a five-pad (with offset) TPJBs provided by an industrial partner. The bearings had a 280 mm inner diameter and were tested in the load between pads configuration. No additional data will be given about the bearings and their characteristics due to a non-disclosure agreement with the company.

Two main types of test are usually carried out with the test apparatus:

1. The bump test to identify the bearing clearance, center, and alignment
2. The dynamic identification test to identify bearing stiffness and damping coefficients.

In the first test a rotating force vector applied to the bearing housing is generated by sinusoidal signals with 90° phase difference. The oil is not supplied, and the shaft is not rotating. The force must be increased until the polygon shaped orbit does not change.

In the second test the forces applied during excitation can contain one ("single tone" test) or more ("multitone" test) frequency components. As dynamic coefficients can vary with frequency, excitation tones are chosen below and above the rotational frequency avoiding harmonics and test bench resonances. The required synchronous values are then obtained by interpolation to avoid imbalance disturbance. Multitone tests with n frequency components have been proven to provide the same results of n single tone tests but in the time of a single test. For the identification of the dynamic linear coefficients, two tests with linearly independent excitations are required for each excitation frequency. The two tests consist in two distinct excitations, vertical (subscript y) and horizontal (subscript x), respectively, obtained using the dynamic actuators in in-phase (subscript f) and anti-phase (subscript a) operation modes. The bearing impedance matrix **H**, expressed in terms of stiffness (k) and damping (c) coefficients:

$$\begin{bmatrix} H_{xx} & H_{xy} \\ H_{yx} & H_{yy} \end{bmatrix} = \begin{bmatrix} k_{xx} & k_{xy} \\ k_{yx} & k_{yy} \end{bmatrix} + i\omega \begin{bmatrix} c_{xx} & c_{xy} \\ c_{yx} & c_{yy} \end{bmatrix} \tag{1}$$

where ω is the excitation frequency, is determined in the frequency domain, using the Fast Fourier Transform (FFT), by multiplying the [2 × 2] bearing force complex matrix by the corresponding inverse displacement complex matrix:

$$\begin{bmatrix} H_{xx} & H_{xy} \\ H_{yx} & H_{yy} \end{bmatrix} = \begin{bmatrix} F_{bxf} & F_{bxa} \\ F_{byf} & F_{bya} \end{bmatrix} \begin{bmatrix} D_{xf} & D_{xa} \\ D_{yf} & D_{ya} \end{bmatrix}^{-1} \tag{2}$$

where F_b indicates the amplitude of the force transform and D indicates the amplitude of the displacement transform. Synchronous coefficients are obtained at the shaft rotational frequency.

The stiffness dynamic coefficients have shown to be practically independent of frequency for a wide range starting from quasi-static conditions. They are the basis of the simplified models that will be presented in Section 3.2.

In addition to such tests, in order to ascertain the linear and nonlinear range of displacements prior to the dynamic characterization campaign, and also to have a reference stiffness estimate, another procedure has been developed. The stator is subjected to a slowly rotating force vector generated by equal amplitude sinusoidal forces with a 90° phase angle difference, applied by the two dynamic actuators. The force low rotational speed allows to consider the damping coefficients negligible. This work focuses on this latter test procedure that makes it possible to evidence nonlinear effects by detecting the loss of the typical elliptical orbit related to linear bearing film stiffness. In the tests reported in this work the shaft rotational speed was set at 1000 rpm while the frequency of the rotating load was set at 1 round every 100 s (i.e., 0.01Hz). Different static load levels were applied in load-between-pad configuration combined with different dynamic load levels ranging 3% to 36% of the static load. Forces and displacements were recorded at a rate of one sample/s.

3. Results

3.1. Experimental Results

Figure 4 shows the plots of the applied horizontal and vertical dynamic load, obtained in four different tests for the four-pad TPJB for two levels of static load, with the higher (L2) about double the lower one (L1).

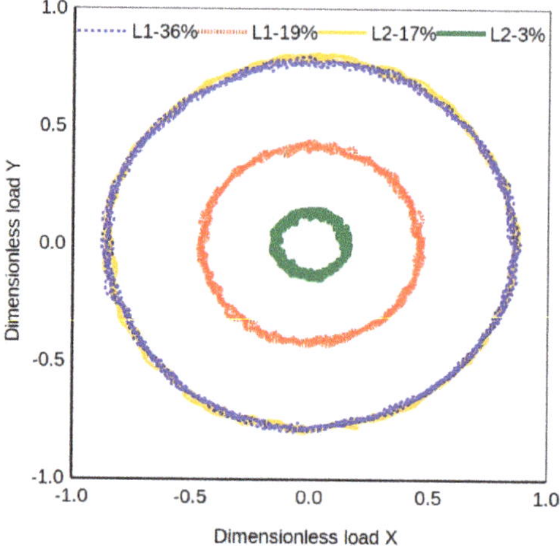

Figure 4. Horizontal and vertical dynamic load plot for the four-pad tilting pad journal bearings (TPJB). Red and blue lines for the lower static load L1 and increasing dynamic/static load ratio, and green and yellow for the higher static load L2 and increasing dynamic/static load ratio.

The dynamic load plot is circular because it is a rotating vector with a constant amplitude controlled by means of the dynamic actuators with the feedback of the load sensor system. The dynamic load is indicated in the label as percentage of the static load. About five rotating load cycles are shown for each test. The corresponding stator orbits are shown in Figure 5. The zero values correspond to the central position of the bearing. The displacement fluctuation, related to the centrifugal force due to the shaft rotation, is noticeable particularly for the low static load conditions. It is evident that the orbit position is related to the static load level while its shape is greatly influenced by the dynamic/static load ratio, becoming more elliptical and closer to the linear orbits (Figure 5 case L2-3%) as the ratio decreases. The shapes are similar to the ones theoretically found by direct integration of the Reynolds equation in References [9,11,12] for four-pad TPJBs.

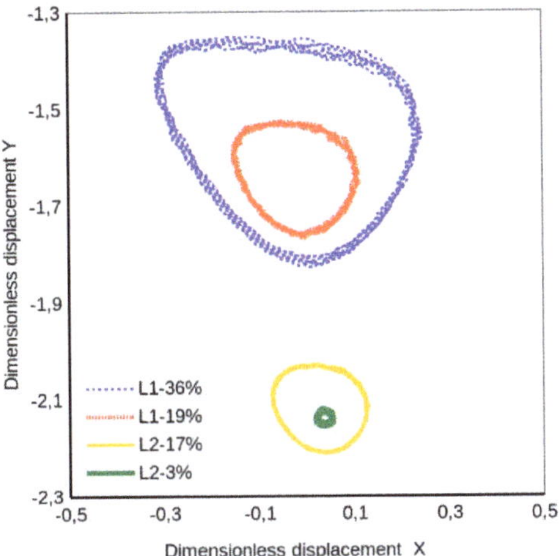

Figure 5. Orbits generated by the rotating force for the four-pad TPJB. Red and blue lines for the lower static load L1 and increasing dynamic/static load ratio, and green and yellow for the higher static load L2 and increasing dynamic/static load ratio.

3.2. Simple Analytical Models

In order to give a possible explanation for the obtained orbits, the influence of different factors was evaluated by simulation with simple bearing models represented by linear and quadratic dynamic coefficients.

Analytical models with increasing complexity were devised assuming hydrodynamic forces having the following non-linear relations with displacements, obtained by adding quadratic terms to the commonly used linear relations, neglecting damping due to the low excitation frequency of the rotating load:

$$f_x = k_{xx}d_x + k_{xy}d_y + k_{xx2}d_x^2 + k_{xy2}d_y^2, \tag{3}$$

$$f_y = k_{yx}d_x + k_{yy}d_y + k_{yx2}d_x^2 + k_{yy2}d_y^2 \tag{4}$$

The main objective of this work was not the identification of nonlinear coefficients but to preliminarily investigate the possibility of replicating the experimental nonlinear orbits with simple nonlinear models. Thus, the linear coefficients of Equations (3) and (4) were set equal to those obtained by the identification tests described in the previous section while the quadratic coefficients were determined by a trial and error procedure to get a good fit of the experimental orbits with analytical ones based on the proposed models.

The simplest analytical linear model takes into account only the direct stiffness coefficients in the linear relation with displacements:

$$d_x = \frac{f_x}{k_{xx}}, \; d_y = \frac{f_y}{k_{yy}}. \tag{5}$$

The linear stiffness model takes into account both direct and cross-coupled stiffness coefficients in the linear relation with displacements:

$$d_x = \frac{k_{yy}f_x - k_{xy}f_y}{k_{xx}k_{yy} - k_{xy}k_{yx}}, \; d_y = \frac{k_{xx}f_y - k_{yx}f_x}{k_{xx}k_{yy} - k_{xy}k_{yx}}. \tag{6}$$

The simplest nonlinear model takes into account only direct linear and quadratic stiffness coefficients:

$$d_x = \frac{-k_{xx} + \sqrt{k_{xx}^2 + 4k_{xx2}f_x}}{2k_{xx2}}, \; d_y = \frac{-k_{yy} + \sqrt{k_{yy}^2 + 4k_{yy2}f_y}}{2k_{yy2}} \tag{7}$$

A second non-linear model takes into account linear direct and cross-coupled stiffness coefficients and quadratic direct stiffness coefficients. First of all, the force and displacement components of Equations (1) and (2) were expressed in polar coordinates:

$$f_x = F\cos(\theta + \varphi), \; f_y = F\sin(\theta + \varphi), \tag{8}$$

$$d_x = R\cos(\theta), \; d_y = R\sin(\theta). \tag{9}$$

After squaring the terms of both Equations (3) and (4), they were summed obtaining an equation of the fourth degree of R that can be solved analytically as a function of θ:

$$\begin{aligned} R^4\left(k_{xx2}^2\cos^4(\theta) + k_{yy2}^2\sin^4(\theta)\right) & \\ +R^3\left[2k_{xx2}\cos^2(\theta)\left(k_{xx}\cos(\theta) + k_{xy}\sin(\theta)\right) + 2k_{yy2}\sin^2(\theta)\left(k_{yx}\cos(\theta) + k_{yy}\sin(\theta)\right)\right] & \\ +R^2\left[\left(k_{xx}\cos(\theta) + k_{xy}\sin(\theta)\right)^2 + \left(k_{yx}\cos(\theta) + k_{yy}\sin(\theta)\right)^2\right] - F^2 = 0. & \end{aligned} \tag{10}$$

A rotating load was imposed as a sum of two sinusoidal functions with the same amplitude and a 90° phase shift, and the corresponding displacements were calculated. The case L1-36% (low static load, high load ratio) was chosen as reference case due to its extreme dynamic load conditions. For the sake of comparison, the same experimentally identified stiffness coefficients for the four-pad TPJB were used in all the analytical models while, for the quadratic coefficients, values tuned to fit the experimental orbits were adopted for the x and y directions for all nonlinear models. The value zero in the diagrams corresponds to the static equilibrium position. Further non-linear models were obtained by considering the dependence of stiffness coefficients from the actual vertical load, sum of the static load, and the vertical component of the rotating load, according to a fit of experimental results obtained for different static vertical loads and small dynamic ones. In such a case, a phase angle of a few degrees between the displacement vector and the force vector was included, affecting the model related to Equation (10). The orbits obtained with the different analytical models are presented in Figure 6, with dashed blue lines representing those of the constant stiffness models and red lines representing those of the load dependent stiffness models.

Comparing the orbits of Figure 6, one can observe that if only direct stiffness coefficients are included the orbits are necessarily circular due to the same bearing stiffness along the orthogonal directions. The inclusion of the linear cross-coupled stiffness coefficients in the model modifies the orbit shape in a tilted slightly elliptical one. Including second order direct stiffness coefficients in the model yields an orbit with a shape more similar to the experimental one with three lobes. It seems that including cross-coupled stiffness coefficients in this latter model makes it more adaptable but does not bring significant improvements unless a specific optimization procedure is performed. The inclusion of coefficient load dependence produces a vertical shift of all the orbits, an increase of the ellipticity for the simpler models, and more pronounced lobes for the nonlinear ones.

3.3. Comparison of Analytical and Experimental Results

In order to evaluate the capability of analytical models to simulate the actual bearing behavior, experimental loads and corresponding linear stiffness coefficients were implemented in the models, tuning the quadratic coefficients, to obtain orbits to be compared with the experimental ones. Figure 7 shows a comparison of orbits calculated with different load dependent stiffness models and experimental ones for three different load ratios.

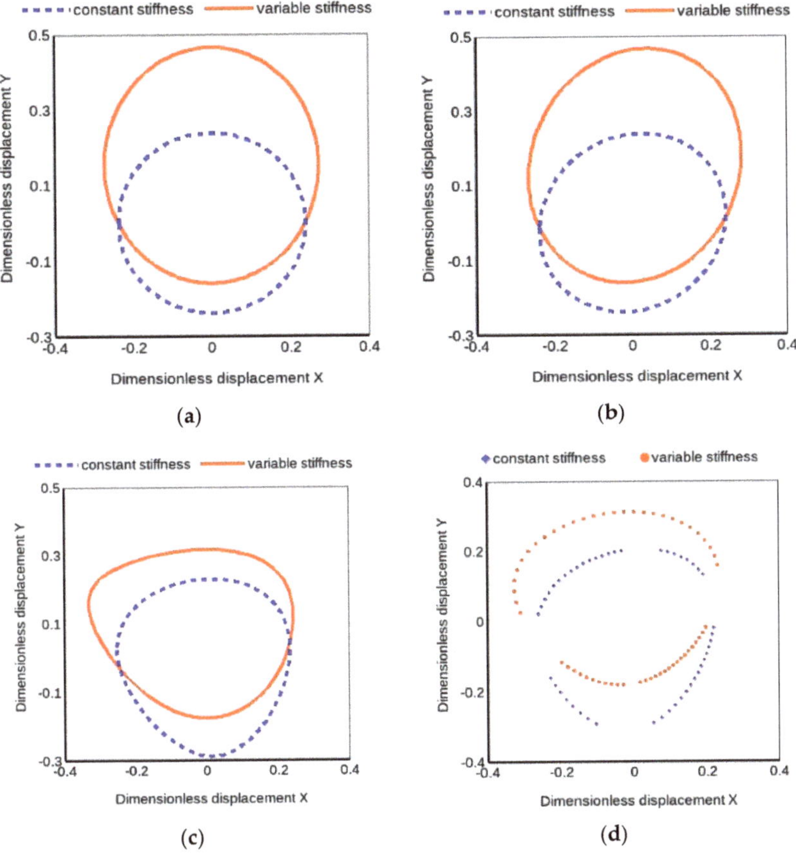

Figure 6. Calculated orbits for the four-pad TPJB for four different models: (**a**) linear constant and load dependent direct stiffness coefficients only; (**b**) linear constant and load dependent direct and cross-coupled stiffness coefficients; (**c**) nonlinear constant and load dependent direct stiffness coefficients only; (**d**) nonlinear constant and load dependent direct stiffness coefficients with linear constant and load dependent cross-coupled stiffness coefficients. Case L1-36%.

The more complex nonlinear model is omitted at this point because its optimization deserves a further in-depth analysis and thus it is left to future development. It is evident especially at high load ratios that the nonlinear model overcomes the limitations of the linear ones and succeeds in replicating the experimental orbit shape. At lower load ratios the differences are less marked as the orbits tend to a more elliptical shape.

For very low load ratios, that is for reduced values of the dynamic force the orbit tends to become more circular and all models, linear and nonlinear, give similar results, as shown in Figure 7c for the case L2-3%, thus indicating a linear behavior. Note that in this case, since both displacement and force experimental values are very small, the measurement relative error increases and it is also quite difficult to control the rotating force vector at such load ratios. Those are the main reasons for the discrepancy of experimental and analytical orbits shown in Figure 7c.

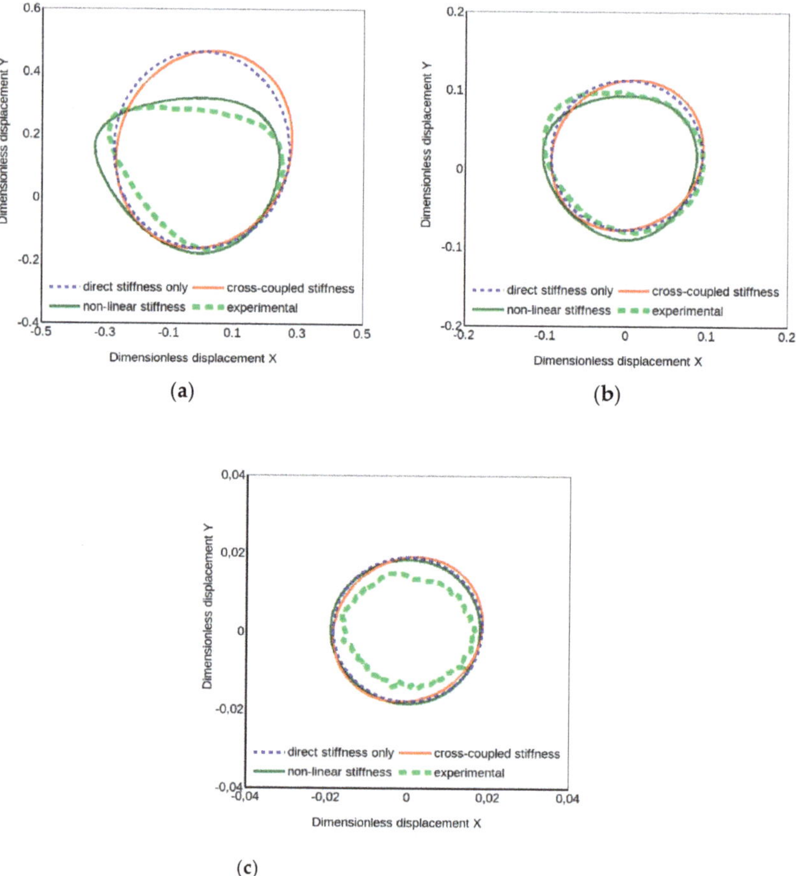

Figure 7. Calculated and experimental orbits of the four-pad TPJB for three different models with load dependent direct stiffness coefficients and two loads (L2>L1): (**a**) L1-36%, (**b**) L2-17%, (**c**) L2-3%.

4. Discussion

The results obtained for the four-pad TPJB that has equal direct stiffness coefficients and equal cross-coupled ones indicate that only models including second order direct stiffness coefficients can replicate the characteristic shape of the experimental orbit.

The non-elliptical shape of the orbits is reflected in the appearance of multiple frequency peaks in the Fast Fourier Transform (FFT) of the displacement signal as shown in Figure 8a (case L1-36%). As the excitation is a single tone force, the FFT content of the displacement with multiple frequencies indicates non-linear or coupled phenomena. Figure 8b shows the results of the same analysis performed for the L2-3% case whose orbit is surely more elliptical (Figure 7c). Note that the shaft rotational frequency (16.67 Hz) is not present in these diagrams focused on the low frequency zone. Figure 9 shows the experimental orbit of the case L1-36% compared with the orbits obtained filtering the results at the force rotational frequency (1X) and twice (2X) and three times (3X) the fundamental frequency. While the 1X harmonic component of displacement corresponds indeed to the linear orbit for low load ratios (case L2-3%, Figure 7c), the 1X filtered ellipse observed for a higher load ratio (case L1-36%) in Figure 9 appears with a tilt that is not justified with a linear model considering the negligible linear damping and cross-coupled stiffness coefficients, thus indicating the apparent effects of nonlinearity.

Note that all data recorded during the rotation of the force vector are plotted in Figure 9b (four cycles in this case) instead of the averaged values as in Figure 7. This provides better evidence of the signal fluctuations due the difficulties in controlling and measuring low values of forces and displacements as mentioned above.

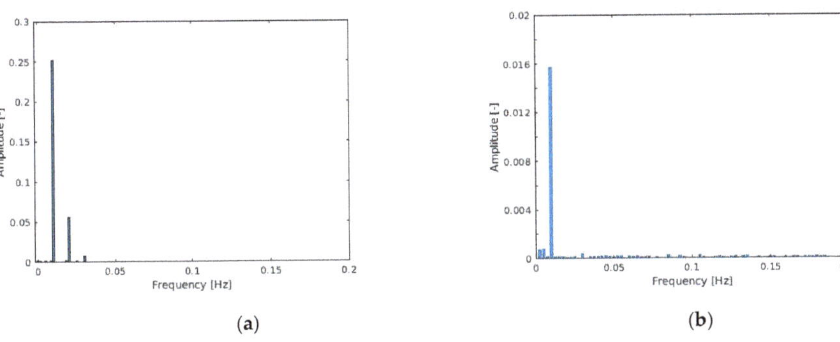

Figure 8. Fast Fourier Transform (FFT) of the X displacement L1-36% (**a**) and L2-3% (**b**) cases.

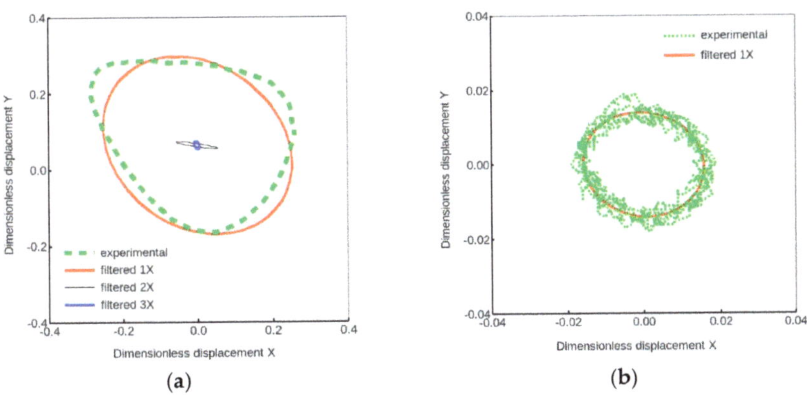

Figure 9. Experimental and filtered orbits for the cases L1-36% (**a**) and L2-3% (**b**).

At the end of the tests some geometrical differences among the pads were also found. It is worth mentioning that geometrical errors can also influence the results, as reported for example in Reference [18].

In order to confirm the previous findings, the case of another TPJB was analyzed. It was quite different from the previous one as it had five offset pads of the same size. It underwent the same tests of the four-pad TPJB in the LBP configuration. Unlike the four-pad TPJB this bearing has quite different direct stiffness coefficients in the x and y directions (k_{yy} greater than k_{xx}), and that has obviously a remarkable impact on the orbit shapes. Figure 10 presents some experimental orbits for the five-pad TPJB for different load ratios and two different static load levels. The load L3 is about 60% of L2, so a little greater than the load L1 used for the four-pad TPJB. Figure 11 presents calculated and experimental orbits for three different models with load dependent direct stiffness coefficients. The difference in direct stiffness coefficients produces the ellipticity even of the orbits of the simpler models but the coefficient load dependence causes a distortion of the ellipse, though there is still a difference in its orientation compared to the experimental one. Again, when the load ratio is small, as in cases L3-5% and L2-3% of Figure 10, the orbit is elliptical and, as shown in Figure 11b, quite close to the classical linear model predicted one. Moreover, the model with quadratic coefficients produces an orbit more similar to the experimental one, particularly noticeable for larger load ratios. Nonetheless there is still margin for an optimized tuning of the estimated quadratic coefficients. Better results can be expected from a best fit optimization involving all dynamic coefficients, including the linear ones. The set of linear coefficients obtained by linear identification could be the starting solution of the nonlinear identification procedure that will be the object of future work.

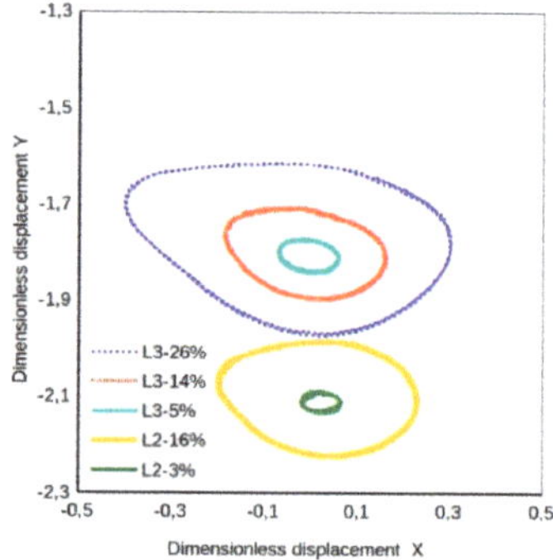

Figure 10. Orbits of the five-pad TPJB generated by the rotating force. Cyan, red, and blue lines for lower static load L3 and increasing dynamic/static load ratio, and green and yellow for higher static load L2 and increasing dynamic/static load ratio.

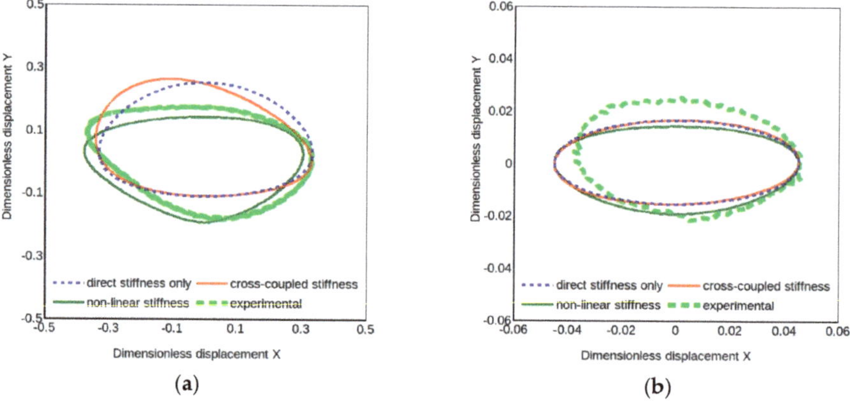

Figure 11. Calculated and experimental orbits of the five-pad TPJB for three different models with load dependent stiffness coefficients and two loads (L3<L2): (**a**) L3-26%, (**b**) L2-3%.

The peculiar three lobe orbit shape, predicted by simulations [9,11,12] for horizontal rotors and found experimentally in the present work, has been ascribed to the number of pads involved in the bearing reaction to the load. With four pads, with a high static load between pads and a rotating load not overcoming the static one, the bearing behavior is similar to that with only two pads, like the one described in Reference [12]. The same behavior occurs with five pads with load between the pads, as can be deduced observing the results reported in this section. 3X components in the journal orbit, in addition to 1X and 2X, have been also reported in Reference [18] for a five-pad TPJB on a floating shaft configuration test rig. Unfortunately, comparison can be only qualitative because experimental conditions are quite different.

5. Conclusions

This paper presented new experimental results on the nonlinear response of large size TPJBs related to the dynamic/static load ratio, showing that nonlinear effects, usually neglected in conventional experimental identification procedures of the bearing dynamic coefficients, should be considered. A quasi-static procedure devised for a preliminary estimate of the bearing stiffness and of the linear displacement range was used to investigate, in a novel way, the nonlinear response of TPJBs. It consists of applying a slowly rotating force to the floating stator. Orbits with particular shapes, different from elliptical, were observed for increasing dynamic to static load ratio.

Numerical simulation using simple bearing models and assuming quadratic stiffness terms and coefficient load dependence generated orbits with shapes similar to the experimental ones for high load ratios where linear models fail, proving the presence of nonlinearities in the bearing reaction to excitation as also indicated by the presence of 2X and 3X harmonic components in the FFT of stator displacements but absent in the FFT of the dynamic load.

These results are the first step for a further study on nonlinear identification of first and higher order coefficients by optimization techniques.

Author Contributions: Conceptualization, E.C. and P.F.; Data curation, E.C. and P.F.; Methodology, E.C. and P.F.; Writing – original draft, E.C. and P.F.; Writing – review & editing, E.C. and P.F.

Funding: This research received no external funding.

Acknowledgments: The authors thank Francesco Maestrale and Matteo Nuti of AM Testing s.r.l. for performing the experimental tests.

References

1. Lund, J.W. Review of the Concept of Dynamic Coefficients for Fluid Film Journal Bearings. *J. Tribol.* **1987**, *109*, 37–41. [CrossRef]
2. Tiwari, R.; Lees, A.W.; Friswell, M.I. Identification of Dynamic Bearing Parameters: A Review. *Shock Vib. Dig.* **2004**, *36*, 99–124. [CrossRef]
3. Dimond, T.W.; Sheth, P.N.; Allaire, P.E.; He, M. Identification methods and test results for tilting pad and fixed geometry journal bearing dynamic coefficients—A review. *Shock Vib.* **2009**, *16*, 13–43. [CrossRef]
4. Qiu, Z.L.; Tieu, A.K. The Effect of Perturbation Amplitudes on Eight Force Coefficients of Journal Bearings. *Tribol. Trans.* **2008**, *39*, 469–475. [CrossRef]
5. Cha, M.; Kuznetsov, E.; Glavatskih, S. A comparative linear and nonlinear dynamic analysis of compliant cylindrical journal bearings. *Mech. Mach. Theory* **2013**, *64*, 80–92. [CrossRef]
6. Zhao, S.X.; Dai, X.D.; Meng, G.; Zhu, J. An experimental study of nonlinear oil-film forces of a journal bearing. *J. Sound Vib.* **2005**, *287*, 827–843. [CrossRef]
7. Meruane, V.; Pascual, R. Identification of nonlinear dynamic coefficients in plain journal bearing. *Tribol. Int.* **2008**, *41*, 743–754. [CrossRef]
8. Vania, A.; Pennacchi, P.; Chatterton, S.; Tanzi, E. Sensitivity Analysis of Non-Linear Forces in Oil-Film Journal Bearings. In Proceedings of the World Tribology Congress 2013, Torino, Italy, 8–13 September 2013; pp. 1–4.
9. Fillon, M.; Desbordes, H.; Frêne, J.; Wai, C.C.H. A Global Approach of Thermal Effects Including Pad Deformations in Tilting-Pad Journal Bearings Submitted to Unbalance Load. *J. Tribol.* **1996**, *118*, 169–174. [CrossRef]
10. Monmousseau, P.; Fillon, M.; Frene, J. Tansient Thermoelastohydrodynamic Study of Tilting-Pad Journal Bearings under Dynamic Loading. In Proceedings of the International Gas Turbine & Aeroengine Congress & Exhibition, Orlando, FL, USA, 2–5 June 1997; pp. 1–6.
11. Cha, M.; Glavatskih, S. Nonlinear dynamic behaviour of vertical and horizontal rotors in compliant liner tilting pad journal bearings: Some design considerations. *Tribol. Int.* **2015**, *82*, 142–152. [CrossRef]
12. Brancati, R.; Rocca, E.; Russo, R. Non-linear stability analysis of a rigid rotor on tilting pad journal bearings. *Tribol. Int.* **1996**, *29*, 571–578. [CrossRef]
13. Pagano, S.; Rocca, E.; Russo, M.; Russo, R. Dynamic behaviour of a tilting-pad journal bearings. *Proc. Inst. Mech. Eng. Part J J. Eng. Tribol.* **1995**, *209*, 275–285. [CrossRef]

14. Brancati, R.; Pagano, S.; Rocca, E.; Russo, M.; Russo, R. Dynamic behaviour of an unloaded rotor in tilting pad journal bearings. In Proceedings of the ISROMAC-7/7th International Symposium on Transport Phenomena and Dynamics of Rotating Machinery, Honolulu, HI, USA, 22–26 February 1998; pp. 482–490, ISBN 0965246930.

15. Asgharifard-Sharabiani, P.; Ahmadian, H. Nonlinear model identification of oil-lubricated tilting pad bearings. *Tribol. Int.* **2015**, *92*, 533–543. [CrossRef]

16. De Falco, D.; della Valle, S.; Pagano, S.; Rocca, E. Rotors on tilting pad journal bearings: The results of theoretical and experimental models. In *Tribology Research: From Model Experiment to Industrial Problem*; Elsevier Science B.V.: Amsterdam, The Netherlands, 2001; pp. 127–137.

17. Pennacchi, P.; Vania, A. Early detection of non-linear effects in fluid-film journal bearings. In Proceedings of the 6th IFToMM International Conference on Rotor Dynamics, Sydney, Australia, 30 September–4 October 2002; pp. 201–208, ISBN 073341963.

18. Dang, P.; Chatterton, S.; Pennacchi, P.; Vania, A.; Cangioli, F. An Experimental Study of Nonlinear Oil-Film Forces in a Tilting-Pad Journal Bearing. In Proceedings of the International Design Engineering Technical Conferences and Computers and Information in Engineering Conference, 27th Conference on Mechanical Vibration and Noise, Boston, MA, USA, 2–5 August 2015. [CrossRef]

19. Ciulli, E.; Forte, P. Nonlinear effects in the dynamic characterization of tilting pad journal bearings. In Advances in Italian Mechanism Science. In Proceedings of the Second International Conference of IFToMM Italy, Cassino, Italy, 29–30 November 2018; Carbone, G., Gasparetto, A., Eds.; Springer: Cham, Switzerland, 2019; pp. 474–481. [CrossRef]

20. Forte, P.; Ciulli, E.; Saba, D. A novel test rig for the dynamic characterization of large size tilting pad journal bearings. *JPCS* **2016**, *744*, 012159. [CrossRef]

21. Ciulli, E.; Forte, P.; Libraschi, M.; Naldi, L.; Nuti, M. Characterization of High-Power Turbomachinery Tilting Pad Journal Bearings: First Results Obtained on a Novel Test Bench. *Lubricants* **2018**, *6*, 4. [CrossRef]

22. Ciulli, E.; Forte, P.; Libraschi, M.; Nuti, M. Set-up of a novel test plant for high power turbomachinery tilting pad journal bearings. *Tribol. Int.* **2018**, *127*, 276–287. [CrossRef]

MDPI
St. Alban-Anlage 66
4052 Basel
Switzerland
Tel. +41 61 683 77 34
Fax +41 61 302 89 18
www.mdpi.com

Machines Editorial Office
E-mail: machines@mdpi.com
www.mdpi.com/journal/machines